GLACIAL GEOLOGY

HOW ICE SHAPES THE LAND

The Changing Earth Series

JON ERICKSON

Facts On File, Inc.
AN INFOBASE HOLDINGS COMPANY

GLACIAL GEOLOGY: HOW ICE SHAPES THE LAND

Facts On File, Inc.
11 Penn Plaza
New York, NY 10001

Library of Congress Cataloging-in-Publication Data

Erickson, Jon, 1948–
Glacial geology: how ice shapes the land / Jon Erickson.
p. cm. — (The changing earth series)
Includes bibliographical references and index.
ISBN 0-8160-3355-2
1. Glacial landforms. 2. Glacial epoch. I. Title. II. Series:
Erickson, Jon, 1948– Changing earth.
GB581.E75 1996 95–35271
551.3' 15—dc20

Facts On File books are available at special discounts when purchased in bulk quantities for businesses, associations, institutions or sales promotions. Please call our Special Sales Department in New York at 212/967-8800 or 800/322-8755.

Text design by Ron Monteleone
Jacket design by Catherine Rincon Hyman

Printed in the United States of America

RRD FOF 10 9 8 7 6 5 4 3 2 1

This book is printed on acid-free paper.

CONTENTS

TABLES

ACKNOWLEDGMENTS

The author thanks the following organizations for providing photographs for this book: the National Aeronautics and Space Administration (NASA), the National Oceanic and Atmospheric Administration (NOAA), the National Park Service, the U.S. Air Force, the U.S. Coast Guard, the U.S. Department of Agriculture–Soil Conservation Service, the U.S. Geological Survey (USGS), and the U.S. Navy.

INTRODUCTION

Many northern regions worldwide owe their unique landscapes to enormous glaciers that swept down from the poles during the last ice age. Glaciation was so widespread ice sheets 2 miles thick enveloped the upper North American and Eurasian continents, Antarctica, and parts of South America and Australia. In addition, mountain glaciers grew on peaks that are now ice free, their legacy the sculpted rock visible in the high ranges of the world. In many areas, the glaciers stripped off entire layers of sediment down to bare bedrock, erasing the entire geologic history of the region.

Several glaciations have visited the Earth during the last half of its existence, when continents wandered into the cold polar regions. Variations in the Earth's orbital motions also had a major influence on the inception of the ice ages, causing cool summers and perpetual ice fields, allowing winter's snowfall to accumulate into glacial ice. The loss of greenhouse gases, principally carbon dioxide and methane, lowered global temperatures significantly, sustaining the ice ages. Even the warming rays of the sun might have appreciably decreased intensity to initiate global cooling and ice ages.

The power of glacial erosion is well demonstrated by deep-sided valleys carved out of mountain slopes by thick sheets of roving ice. Glacial erosion radically modified the shape of stream channels occupied by glaciers a mile or more thick flowing out of the mountains. Hanging valleys carved by glaciers above the main stream valley often feature magnificent waterfalls. In mountainous coastal regions, glaciers gouged long, narrow, steep-sided inlets that adorn the northern coastlines.

Much of the world's northern terrain is dotted with glacial lakes excavated by moving glaciers. Lakes also formed by the melting of large blocks of ice buried under glacial outwash debris. When the great ice sheets melted at the end of the last ice age, about 12,000 years ago, massive floods raged across the continents and scoured the surface, leaving a desolate landscape of ruptured ground. Outwash streams from glacial meltwater also played a major role in carving out the landscape. Long, winding sand deposits were formed by glacial outwash streams running through tunnels beneath the ice sheets. Unusual glacial hillocks that still defy explanation are peppered throughout the northlands.

In the alpine regions, glaciers gouged out large pits as they flowed down mountain peaks. The glaciers extended far down the valleys, grinding rocks on the valley floors as they advanced and receded. Rivers of solid ice with rocks embedded in them ground down the valley floors, like a giant file. As the advancing glaciers sliced down the mountainsides they left their marks on the valley floors, including large areas of polished and deeply furrowed rocks and huge heaps of rocks, along with other unusual landforms left behind when the glaciers melted and retreated back to the poles, where they sit, poised, waiting for the next onslaught of ice.

1

DISCOVERING THE ICE AGES

Historically, prophets from many cultures have foretold of the destruction of the Earth by fire or flood, but no mention has ever been made about ice, though an old Norse legend holds that a period of endless winter occurred when seas froze solid. Such a tale of disaster might have originated out of distant memories of a great ice age, when thick sheets of ice swept down from the polar regions and covered much of the northern lands. Only in the last two centuries have scientists begun to uncover the geologic clues leading to the discovery of worldwide glaciation.

MILESTONES IN GEOLOGY

Around A.D. 1500, the Italian artist and scientist Leonardo da Vinci claimed that fossils were the remains of once-living organisms and not inorganic substances as was previously thought. He also believed the presence of fossil seashells in the mountains was evidence that the distribution of lands and seas had changed through time.

The 17th-century Danish physician and geologist Nicolaus Steno recognized that in a sequence of layered rocks undeformed by folding or faulting, each bed formed after the one below it and before the one above it. Steno's law of superposition might seem obvious today, but during his time it was hailed as an important scientific discovery. Steno also presented the principle of original horizontality, which states that sedimentary rocks were originally laid down in the ocean horizontally and subsequent folding and faulting uplifted them out of the sea and inclined them at steep angles.

When horizontal rocks overlie angled rocks, they represent a gap in geologic time known as an angular unconformity (Fig. 1–1). Furthermore, if a body of rock cuts across the boundaries of other rock units, it must be younger than those it intercepts. This cross-cutting relationship principle holds that granitic intrusions are younger than the rocks they invade. A sequence of rocks placed in their proper sequence without folding or faulting is called a stratigraphic cross section.

The Scottish geologist James Hutton, known as the father of geology, put forward the theory of uniformitarianism, also called gradualism, in 1785. Simply stated, it means the present is the key to the past. In other words, the forces that shaped the Earth are uniform and operate in a similar manner and at the same rate today as they did throughout geologic history. Therefore, present events have counterparts in rocks laid down a very long time ago.

Hutton believed contemporary rocks at the surface were formed by the waste of older rocks that were laid down in the sea, consolidated under

Figure 1–1 Quarternary terrace deposits resting unconformably on eroded edges of Monterey shale, San Luis Obispo County, California. Photo by G. W. Stose, courtesy of USGS

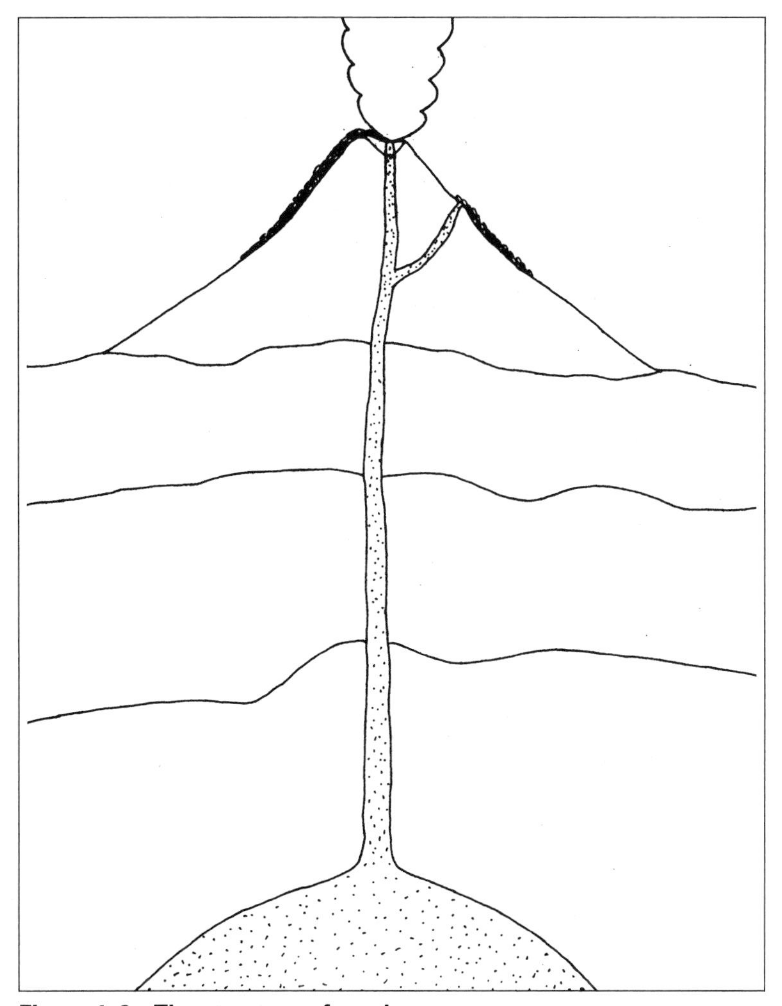

Figure 1–2 The structure of a volcano.

great pressure, and upheaved by the expanding power of the Earth's subterranean heat. He envisioned the planet's own internal heat as the prime mover behind these slow changes. From these observations, Hutton advanced the theory of plutonism, named for Pluto, the Greek god of the underworld. The theory held that the depths of the Earth are in a constant state of turmoil, and that molten matter rises to the surface through cracks or fissures, and causes volcanoes to erupt (Fig. 1–2).

Geologists had long ago recognized that rocks were molten in the Earth's interior and manifested themselves as volcanoes. Moreover, temperatures in deep mines increased with depth, indicating that rocks grew progressively hotter toward the center of the planet. Hutton called this phenomenon the "great heat engine." He believed the heat was left over from when the Earth formed, from its original molten state.

The British geologist Sir Charles Lyell, born in 1797, the year of Hutton's death, took up where his predecessor left off and did much to bring about worldwide acceptance of the theory of uniformitarianism. Lyell gathered many observations about rocks and landforms in western Europe, showing they were the products of the same processes in existence today, only after the passage of immense blocks of time.

Many geologists, however, felt this theory did not adequately explain all the geologic forces at work because some events in the past were not slowly evolving but were comparatively sudden. This opposing view, called catastrophism, had its most ardent supporter in the French geologist Georges Cuvier (1769–1832), who believed the Earth's geologic history was a series of catastrophes.

Adherents pointed to gaps in the geologic record and the extinctions of large numbers of species. Geologists thought the Earth underwent periods of catastrophic death of all life, after which it began anew. This explained the abundance of fossils at certain stages in the geologic record. By the 1700s, most geologists began to accept fossils as the remains of organisms (Fig. 1–3) because they closely resembled living creatures.

Although the existence of fossils has been known since the early Greeks, it was not until the late 18th century that their significance as a geologic tool was discovered. During the 1790s, the English civil engineer William Smith, known as the father of English geology, found that rock formations

Figure 1–3 Fossil brachiopod casts. Photo by E. B. Hardin, courtesy of USGS

in canals he helped build across Great Britain contained fossils different from those in beds below or above.

Smith also noticed that sedimentary strata in widely separated areas could be identified by their distinctive fossil content. He suggested that two rock units from different localities could be regarded as the same age provided they contained identical fossils. Therefore, sedimentary strata in widely separated areas could be identified by their distinctive fossil content.

In the early 1800s, the French geologists Georges Cuvier and Alexandre Brongniart found that certain fossils in rocks around Paris were restricted to specific beds. The geologists arranged fossils in a chronological order and discovered they varied in a systematic way according to their positions in the rock formations. Fossils in the higher beds more closely resembled modern forms than those farther down the geologic column. Also, the fossils did not occur randomly but in a determinable order from simpler to more complex.

These observations led to one of the most important and basic principles of historical geology, whereby segments of geologic time could be identified by their distinctive fossil content. This became the basis for the establishment of the geologic time scale and the beginning of modern geology. The major geologic periods were delineated by 19th-century geologists, mostly in Great Britain and western Europe (Fig. 1–4). The periods were named for localities with the best rock exposures. For example, the Jurassic

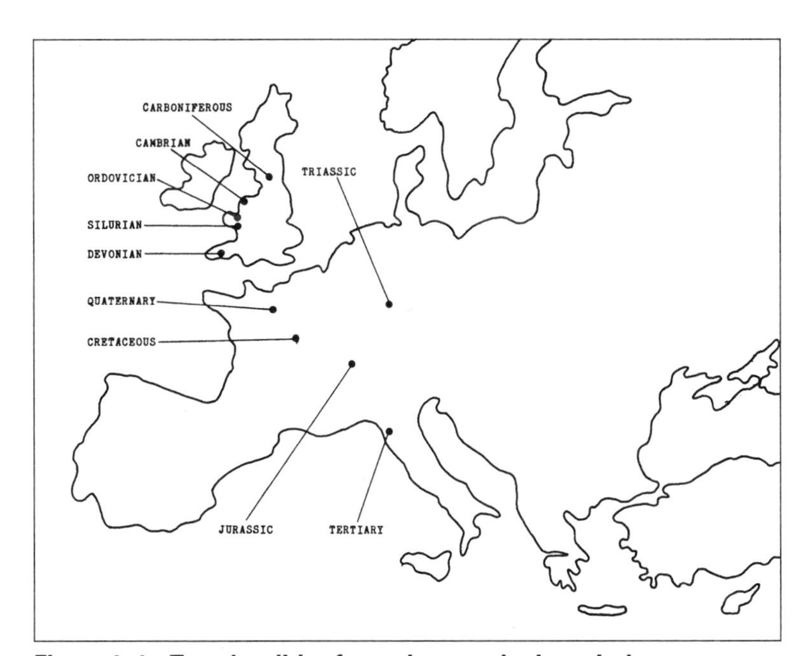

Figure 1–4 Type localities for various geologic periods.

TABLE 1–1 THE GEOLOGIC TIME SCALE

Era	Period	Epoch	Age (millions of years)	First life forms	Geology
		Holocene	0.01		
	Quaternary				
		Pleistocene	3	Man	Ice age
Cenozoic		Pliocene	11	Mastodons	Cascades
		Miocene	26	Saber-toothed tigers	Alps
	Tertiary	Oligocene	37		
		Eocene	54	Whales	
		Paleocene	65	Horses Alligators	Rockies
	Cretaceous		135		
				Birds	Sierra Nevada
Mesozoic	Jurassic		190	Mammals	Altlantic
				Dinosaurs	
	Triassic		250		
	Permian		280	Reptiles	Appalachians
		Pennsylvanian	310		Ice age
				Trees	
	Carboniferous				
Paleozoic		Mississippian	345	Amphibians Insects	Pangaea
	Devonian		400	Sharks,	
	Silurian		435	Land plants	Laursia
	Ordovician		500	Fish	
	Cambrian		570	Sea plants	Gondwana
				Shelled animals	
			700	Invertebrates	
Proterozoic			2,500	Metazoans	
			3,500	Earliest life	
Archean			4,000	Oldest rocks	
			4,600	Meteorites	

Figure 1–5 A scientist dating a sample using the radiocarbon method. Courtesy of USGS

period took its named from the Jura Mountains in Switzerland, which feature a well-fossilized limestone formation.

In order to develop a geologic time scale applicable over the entire world, rocks of one locality are correlated or matched with rocks of similar age in another location. By correlating rocks from one place to another over a wide area, a comprehensive view of the geologic history of a region can be obtained. A bed or a series of beds can thus be traced from one outcrop to another by recognizing certain distinctive features in the rocks.

Although these methods were sufficient for tracing rock formations over relatively short distances, they were inadequate for matching rocks over long stretches, such as from one continent to another. Only fossils could be used to correlate rock formations between widely separated areas or between continents.

Because technology for dating rocks did not yet exist, the entire geologic record was compiled using relative dating techniques based on the fossil content of the rocks, which places geological time units in their proper sequence without reference to their actual age. Only during this century, after the development of radiometric dating techniques based on the decay of radioactive isotopes, (Fig. 1–5) have absolute dates been applied to units of geologic time.

The largest divisions of the geologic time scale are eras. They include the Precambrian—the age of prelife, the Paleozoic—the age of ancient life, the Mesozoic—the age of middle life, and the Cenozoic—the age of new life. The eras of the Phanerozoic eon, the last 600 million years, are divided into smaller units called periods. There are seven periods in the Paleozoic, three in the Mesozoic, and two in the Cenozoic.

Periods are marked by fewer profound changes in organisms compared to the eras, which mark the boundaries of mass extinctions, proliferations, or rapid transformations of species. The two periods of the Cenozoic have been subdivided into seven epochs due to the greater detail provided by more recent rocks. The Pleistocene epoch, beginning about 3 million years ago, is unique in geologic history because of a succession of ice ages.

MODERN DISCOVERIES

The British naturalist Charles Darwin did for biology what Hutton and Lyell did for geology. It is interesting to note that Darwin was trained as a geologist, but today he tends to be viewed as a biologist. In 1859, he published *On the Origin of Species,* outlining his theory of evolution. Darwin observed the relationships between animals on islands and on adjacent continents as well as between species and fossils of their extinct relatives. This study led him to the conclusion that species had been continuously evolving through time (Fig. 1–6).

Modern nuclear physics was born near the turn of the 20th century with many discoveries explaining the mysteries locked up inside the atom. In 1879, William Crookes produced ionization in a gas with an electric discharge. Wilhelm Roentgen discovered X rays while experimenting with an electrical discharge tube in 1895. A year later, Henri Becquerel found similar rays, called gamma rays, emanating from uranium. In 1897, J.J. Thomson identified the electron, the negatively charged particle orbiting the atom. In 1898, Ernest Rutherford discovered radioactive alpha and beta particles, and Pierre and Marie Curie isolated radium and other radioactive elements. Some 50 years later, radioactive dating techniques were developed to accurately date geologic events.

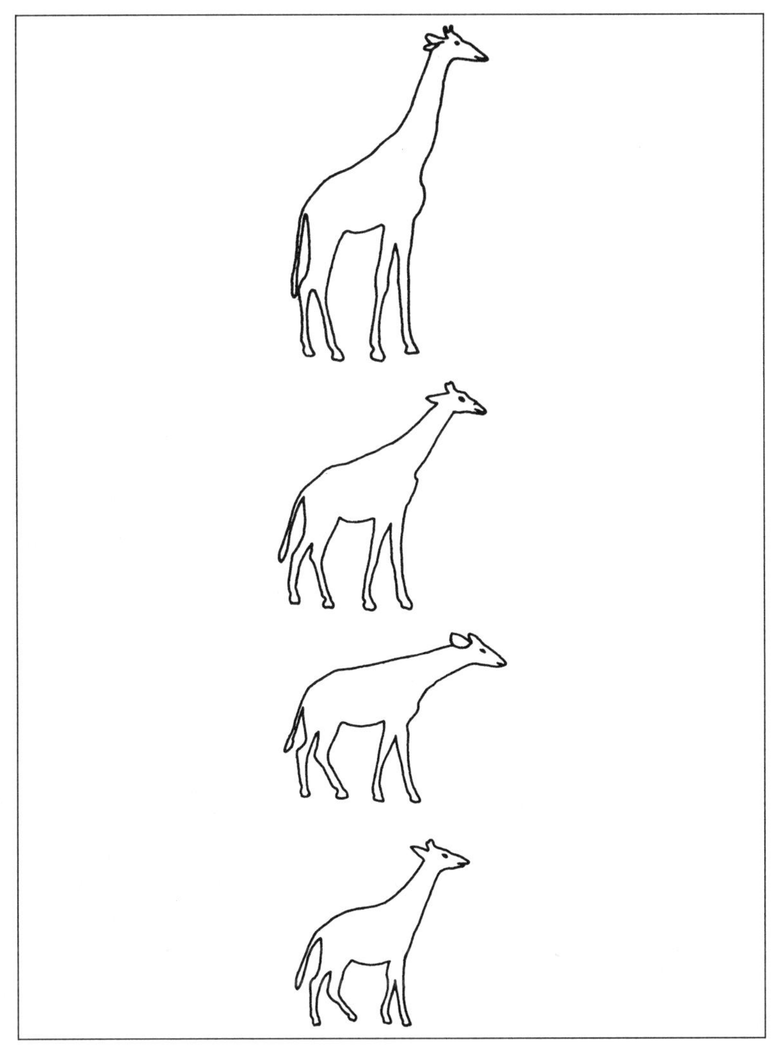

Figure 1–6 Evolution of the giraffe in response to dieting on tree leaves.

In 1894, the 19th-century English astronomer Walter Maunder discovered an apparent period of minimal sunspot activity from 1645 to 1715, now known as the Maunder Minimum. He noticed a marked decline in sunspot numbers over the entire 70-year period, when average global temperatures dropped 1 degree Celsius (2 degrees Fahrenheit). It was blamed for a span of unusually cold weather in Europe and North America during the Little Ice Age, a cooling trend that began around 1500 and ended some 350 years later.

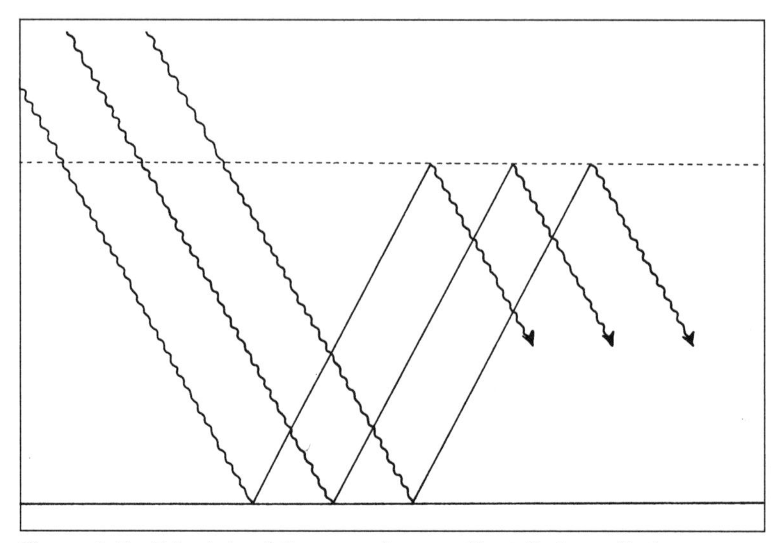

Figure 1–7 Principle of the greenhouse effect. Solar radiation passes through the atmosphere, is converted into infrared radiation on the ground; this radiation escapes upward, is absorbed by greenhouse gases, and is then reradiated toward the surface.

The mechanics of the greenhouse effect (Fig. 1–7) were recognized before the turn of the 20th century, when, in 1896, the Swedish chemist Svante Arrhenius predicted the effects of atmospheric carbon dioxide on the climate. He concluded that past glacial periods might have occurred largely because of a reduction of carbon dioxide in the atmosphere. Arrhenius also estimated that a doubling of the concentration of atmospheric carbon dioxide would cause a global warming of about 5 degrees Celsius (9 degrees Fahrenheit) surprisingly aligned with present-day greenhouse models.

In 1909, the Yugoslavian seismologist Andrija Mohorovicic discovered the boundary between the mantle and the crust, now known as the Mohorovicic discontinuity, or simply the Moho. By studying seismic waves from earthquakes, seismologists could determine certain properties of the Earth's interior, like a geologic X ray. They established that the core had an inner solid portion and an outer liquid layer. The core was surrounded by a semisolid mantle, which was covered by a thin crust, giving the Earth a structure similar to an egg.

The early 20th-century German meteorologist Alfred Wegener noticed a correspondence between the shapes of continental coastlines on either side of the Atlantic Ocean, the similarity between geologic provinces (Fig. 1–8), and the resemblance of fossils in South America and Africa. Matching mountain ranges with similar rock formations also existed on opposite continents. Even the ancient climatic conditions were remarkably the same.

In 1915, Wegener contended that some 200 million years ago all landmasses combined into a single large continent named Pangaea, from Greek meaning "all lands." The rest of the world consisted of a large ocean called Panthalassa, from Greek meaning "universal sea." The continents rifted apart during the Jurassic period and scattered to their present geographic locations. The breakup and collisions of continents appropriately explained the formation of mountain ranges, the evolution and extinction of species, and the climatic conditions of the past.

Recently, unrefutable evidence for seafloor spreading and subduction (Fig. 1–9) has led to an entirely new conceptual model of the Earth known as the theory of plate tectonics. The position of the continents and the creation of landforms could now be explained by the interaction of several lithospheric plates that make up the Earth's crust. When two plates col-

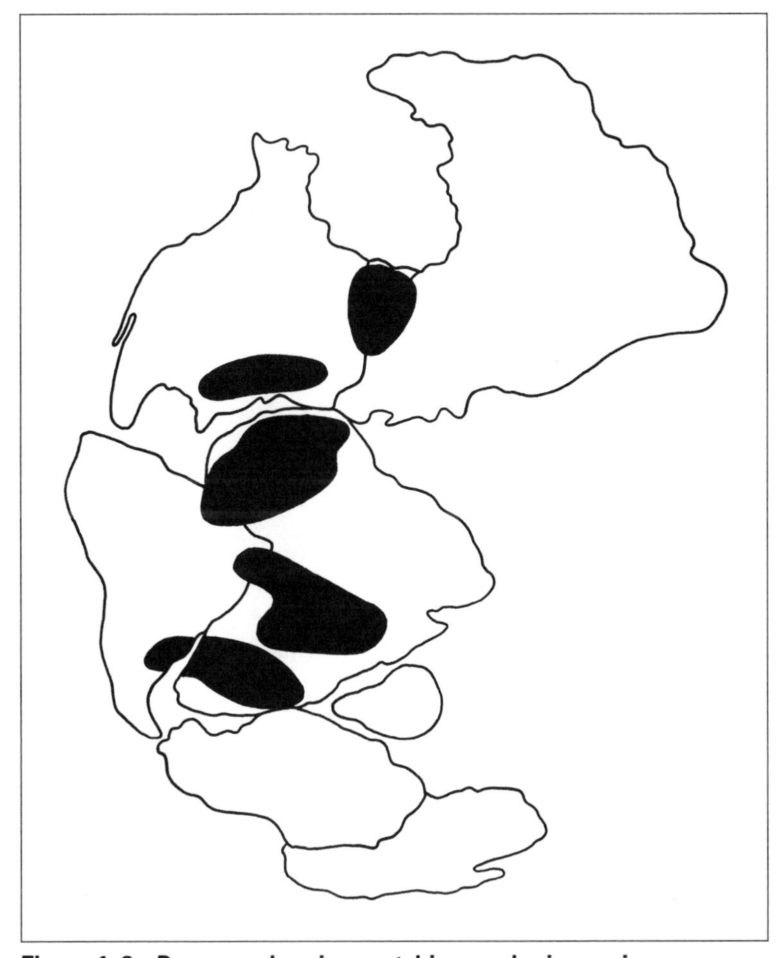

Figure 1–8 Pangaea showing matching geologic provinces.

TABLE 1–2 RADIATION AND EXTINCTION OF SPECIES

Organism	Radiation	Extinction
Mammals	Paleocene	Pleistocene
Reptiles	Permian	Upper Cretaceous
Amphibians	Pennsylvanian	Permian-Triassic
Insects	Upper Paleozoic	(none known)
Land plants	Devonian	Permian
Fishes	Devonian	Pennsylvanian
Crinoids	Ordovician	Upper Permian
Trilobites	Cambrian	Carboniferous & Permian
Ammonoids	Devonian	Upper Cretaceous
Nautiloids	Ordovician	Mississippian
Brachiopods	Ordovician	Devonian & Carboniferous
Graptolites	Ordovician	Silurian & Devonian
Foraminiferans	Silurian	Permian & Triassic
Marine invertebrates	Lower Paleozoic	Permian

lided, mountains rose; when one plate overrode one another, volcanoes erupted; and when two plates slid past each other, earthquakes shattered the land.

During the 1920s and 1930s, the Serbian astronomer Milutin Milankovitch calculated the changes of incoming solar radiation for every latitude during all seasons. He discovered that the three orbital cycles—the shape of the Earth's orbit and the tilt and wobble of its rotational axis—coincided with the 100,000-, 41,000-, and 23,000-year ice age cycles. Milankovitch suggested these three factors worked together to vary the amount of sunshine reaching the higher latitudes.

In 1941, Milankovitch put forward the theory of orbital cycles, which stated that climatic changes brought about by variations in the Earth's orbital motions were responsible for triggering the ice ages. The astronomical theory of the ice ages was first suggested by the French mathematician Joseph Adhemar in 1842, just a few years after the Swiss geologist Louis Agassiz proposed his ice age theory.

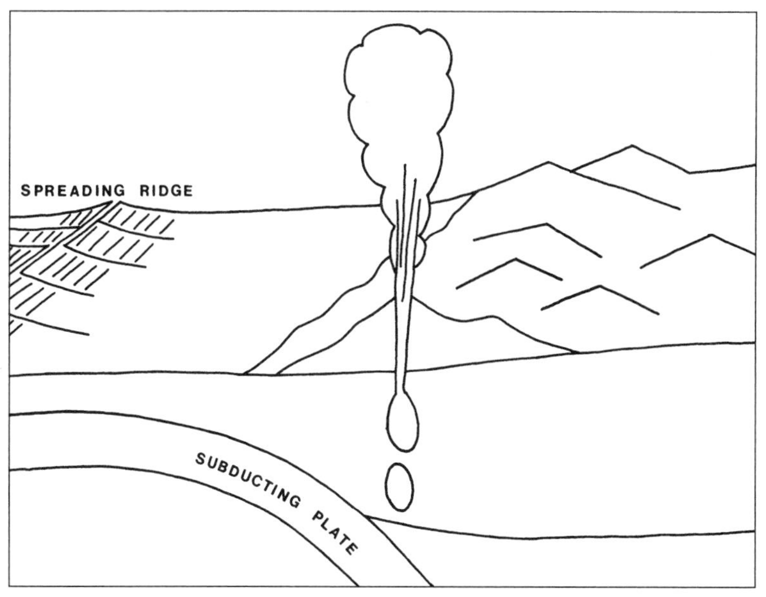

Figure 1–9 Evidence of seafloor spreading and subduction led to the theory of plate tectonics.

BIRTH OF GLACIOLOGY

Geologists of the late 18th century often puzzled over huge blocks of granite weighing several thousand tons called erratic glacial boulders, which were strewn across the mountainous regions in many northern lands (Fig. 1–10). Boulders lying in the Jura Mountains could be traced back to the Swiss Alps over 50 miles away. Many geologists argued that the boulders were swept out of the mountains by the "Great Flood," which also might have divided the continents with opposing shorelines as though they were wide riverbanks.

In 1760, the Swiss geologist Horace de Saussure noticed that downstream from the foot of a glacier the surfaces of projecting rocks along the glacial valley floor looked strikingly different from those high up on the sides of the valley. The higher rocks were rough and jagged, while those lower down were rounded, smooth, and covered with parallel scratches pointing down the valley. Rocks and boulders lay scattered about as though simply dumped there. These observations led de Saussure to conclude that glaciers had once extended far down the valley, grinding the rocks on the valley floor as the ice advanced and retreated.

In 1795, James Hutton described the Alps as once having been covered by a mass of ice, and immense glaciers carried blocks of granite for long distances. Most geologists, however, refuted the idea that a river of solid ice with rocks imbedded in it traveled along the valley floor, grinding down the rocks as the glacier flowed over them like a gigantic file. They refused to believe that glaciers were as widespread as Hutton predicted and could not conceive of how such glaciers could drop isolated blocks of granite in the most unlikely places.

When the Swiss civil engineer Ignatz Venetz heard accounts of marks left on valley floors by glaciers, he visited several of them in various parts of the Swiss Alps. He investigated grooves on valley floors gouged out by advancing glaciers as they ground their way down the mountainsides. By 1829, Venetz had amassed enough information to theorize that alpine glaciers once covered the Jura Mountains and extended far onto the European plain. These observations convinced him that continental-sized glaciers once covered much of the northern lands, which accounted for their many strange erosional features.

During the 1830s, Charles Lyell suggested that fluctuations in ice cover on the land caused changes in sea level. Water evaporating from the ocean landed on the glaciers as snow, which accumulated into thick ice sheets. Because the water did not return to the sea, the level of the ocean lowered appreciably. Lyell was a strong supporter of Hutton's theory of uniformitarianism, though the theory did not seem to account for the occasional ice ages.

Figure 1–10 A glacial boulder split by frost on moraine above Wright Creek, Sequoia National Park, Tulare County, California. Photo by F. E. Matthes, courtesy of USGS

Figure 1–11 A moraine deposit and associated features on the slope of Red Mountain, Gunnison County, Colorado. Photo by B. H. Bryant, courtesy of USGS

Early in the 19th century, Louis Agassiz compiled the first persuasive evidence that glacial ice once covered large parts of the northern continents. He became the foremost proponent of glacial theory and argued that polished and scratched bedrock, and heaps of sand and rocks in the Swiss Alps, were evidence of a past age of glaciers. Waste material and huge boulders much too large for flowing streams to carry could only have been left by retreating glaciers.

Agassiz realized that landforms and sediments in areas presently free of glaciers (Fig. 1–11) resulted from the action of ice and not by the working of wind and water. He found substantial evidence that glacial ice masses had once blanketed the Swiss Mountains and recognized traces of glaciers throughout the northern parts of Europe, Asia, and North America.

In 1837, Agassiz led an expedition of many prestigious European geologists to the Jura Mountains. On the valley floor, he pointed out large areas of polished and deeply furrowed rocks, miles from any existing glaciers, and heaped rocks called moraines, which marked the farthest extent of former glaciers. Agassiz explained that ice a mile or more thick once buried the valley. The glaciers descended from the mountains and spread across most of northern Europe, destroying everything in their paths.

Agassiz showed how glaciers formed by the slow accumulation of snow in the mountains. The snow crystals on the bottom of the glacier were packed and compressed into solid ice by the weight of the overlying snow. After a great many years, a thick solid sea of ice formed. The ice was viscous and flowed outward or retreated toward the mountains, depending on the prevailing climate. Agassiz discovered additional evidence of massive glaciation in the British Isles, Scandinavia, and the north European plain.

When Agassiz emigrated to the United States, he found that virtually all of the North American continent north of the Ohio and Missouri rivers had once been glaciated. American geologists quickly accepted Agassiz's glacial theories, which explained such strange phenomena as gravel deposits, known as lateral and terminal moraines (Fig. 1–12), along with polished and striated rocks found in the Northeast as well as other parts of the country. Long Island and Cape Cod, along with nearby islands, were discovered to have been built entirely of thick moraines and glacial till (Fig. 1–13). Rocks in western New York State were polished and striated similar to those created by glaciers in the Jura Mountains.

THE ICE AGES

The geologist Lyell originally coined the term Pleistocene to mean a period of recent life based on the fossil record of modern organisms. But since

Figure 1–12 Pitted terminal moraine, Columbia County, Wisconsin.
Photo by W. C. Alden, courtesy of USGS

Figure 1–13 Stratified drift above glacial till, Suffolk County, Long Island, New York. Photo by N. M. Perlmutter, courtesy of USGS

widespread glaciation also occurred during this time, the Pleistocene epoch has since become synonymous with the ice ages. It began about 3 million years ago, which also coincides with the disappearance of certain species of Antarctic algae with a siliceous shell called diatoms. The close of the Pleistocene occurs at the end of the last ice age around 10,000 years ago. The following Holocene epoch, the time of modern life, is equivalent to the Neolithic period in archaeology and marks the beginning of civilization.

The term "ice age" now means a time of maximum glacial extent during the Pleistocene, which witnessed a progression of ice ages, each followed by a short interglacial period similar to the one we live in now. The last ice age began about 115,000 years ago. In North America, the ice was upward of 2 miles thick in places and extended as far south as Oregon and New York. The ice also covered most of Great Britain and northern Europe (Fig. 1–14).

Throughout the world, alpine glaciers enveloped mountains that are currently free of ice. During the height of the ice age, snow lines on mountains in New Guinea, Hawaii, and equatorial Africa lay some 3,000 feet lower than they do today. Thick sheets of ice also covered Antarctica, Greenland, parts of Asia, and western and southern South America.

By the late 1800s, most geologists began to accept Agassiz's glacial theories and soon realized the possibility of more than one episode of glaciation. Successive layers of glacial clay separated by soil or peat

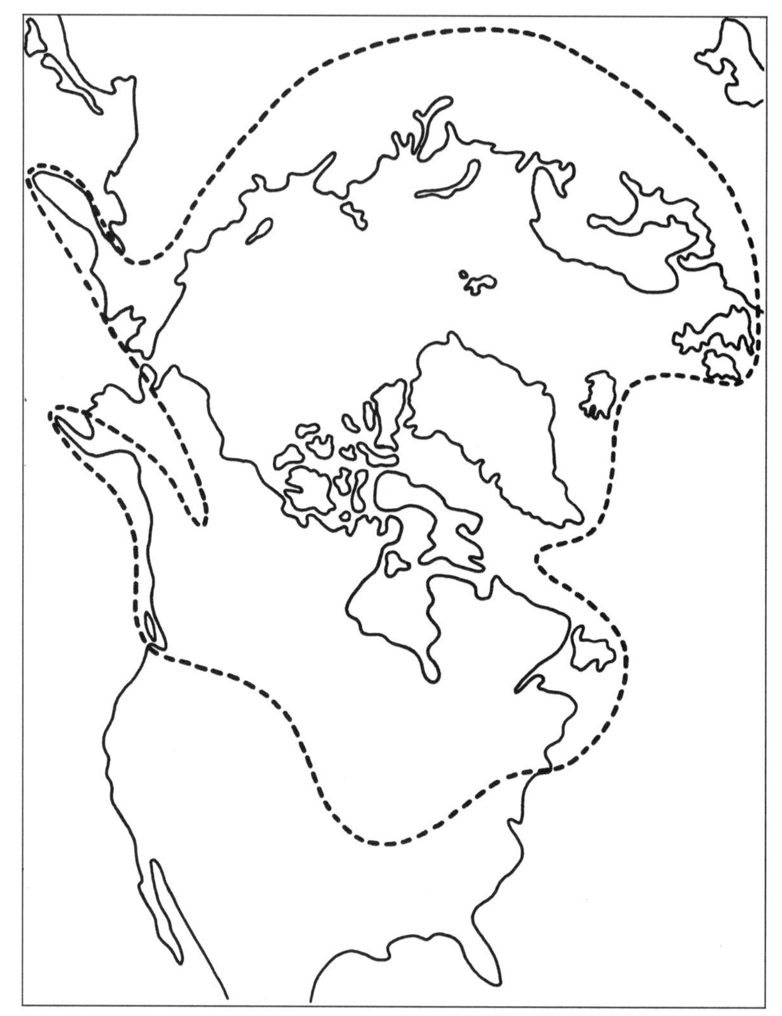

Figure 1–14 The extent of the ice sheets during the last ice age.

suggested that several ice ages followed one after another. In 1909, the Swiss geologists Albrecht Penck and Edward Brukner confirmed the occurrence of at least four separate ice ages during the Pleistocene.

The ice ages in the Alps took their names from Bavarian streams that exposed traces of particular episodes of glaciation. They were named, beginning with the oldest, the Gunz, Mindel, Riss, and Wurm ice ages. In North America, the corresponding episodes of glaciation were given the names of the most affected states and included the Nebraskan, Kansan, Illinoian, and Wisconsin ice ages (Fig. 1–15).

Strong evidence suggested that more than four ice ages punctuated the Pleistocene. Yet many geologists, especially in North America, refused to accept this idea. However, in the early 1950s, the Italian born climatologist Cesare Emiliana, working at the University of Chicago, produced irrefutable evidence for rapid and rhythmic successions of ice ages. But his data came from a seemingly odd place—the bottom of the ocean.

Emiliana obtained his evidence by analyzing the heavy oxygen content of fossil shells of single-celled marine organisms called foraminifera, found in rock cores recovered from the ocean floor. When the forams died, they sank to the bottom and contributed to the seafloor sediments. The carbonate in their shells preserved certain characteristics of the seawater they inhabited, including the differing concentrations of elements such as oxygen.

The ratio of oxygen isotopes (isotopes are identical chemical elements with different atomic masses) in seawater closely follows the proportion of the world's water locked up in the great ice sheets. During a cold climate, the heavier oxygen isotope O_{18} remains behind, as seawater with higher concentrations of the lighter oxygen isotope O_{16} evaporates. The heavier oxygen isotope therefore concentrates in the shells of living organisms. By dating their fossils, Emiliana recognized seven distinct ice ages during the last 700,000 years.

Because of the difficulty in obtaining reliable dates for the ice ages, fully testing glacial theories was difficult because each succeeding ice age tended to erase the evidence of the one before. However, indirect evidence could be secured by studying coral terraces in the tropics. Alternating sea level changes corresponding to the waxing and waning of the ice ages produced a staircaselike structure of coral growth. Radiometric dating techniques

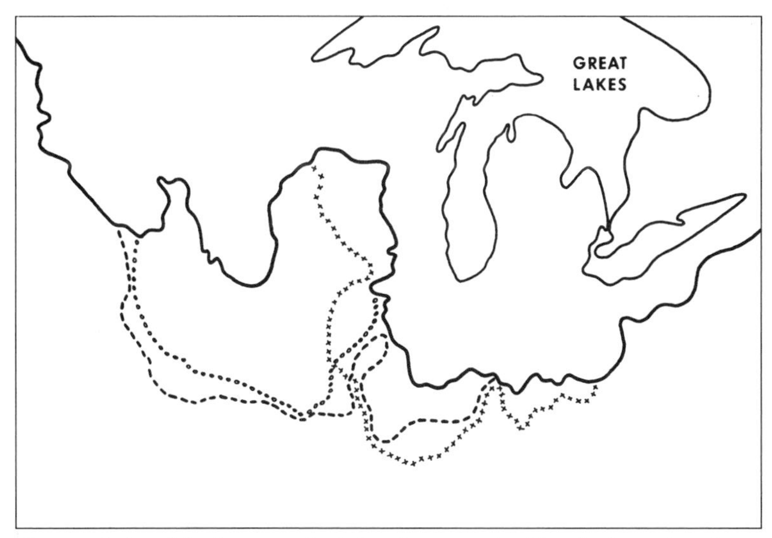

Figure 1–15 Maximum glacial advances in North America from youngest to oldest: Wisconsin (solid line), Illinoian (crosses), Kansan (dashes), and Nebraskan (circles).

determined the ages of the coral terraces, supplying the dates of the glacial events.

An interesting discovery, made entirely by accident, was that Scandinavia and parts of Canada have been slowly rising as much as half an inch a year since the end of the last ice age. Over the centuries, mooring rings on harbor walls in Baltic seaports have risen so far above sea level they could no longer be used to tie up ships. During the last ice age, under the massive weight of the ice, North America and Scandinavia began to sink like overloaded ships. When the ice melted, the crust became more buoyant and began to rise due to the geologic principle of isostasy (Fig. 1–16), which is responsible for maintaining equilibrium in the continental crust.

In Scandinavia, marine fossil beds have risen more than 1,000 feet above sea level since the last ice age. The weight of the ice sheets depressed the continents when the marine deposits were being laid down. When the ice sheets melted, the removal of the weight caused the landmass to rise due to its greater buoyancy. This effect is responsible for maintaining equilibrium in the Earth's crust. Therefore, the lighter continents acted as though they floated on a sea of heavier rocks in the mantle. When the ice sheets melted, the removal of the weight made the land rise.

Geologic evidence suggests that four or five major ice ages occurred during the last 2.4 billion years (Table 1-3). Most of the visible evidence for extensive glaciation comes from deposits of glacial rocks called moraines and tillites. Moraines are unstratified sediments, ranging from sand to boulders. Tillites are a mixture of boulders and pebbles in a clay matrix consolidated into solid rock. Both types of deposits were laid down by glacial ice and exist on every continent.

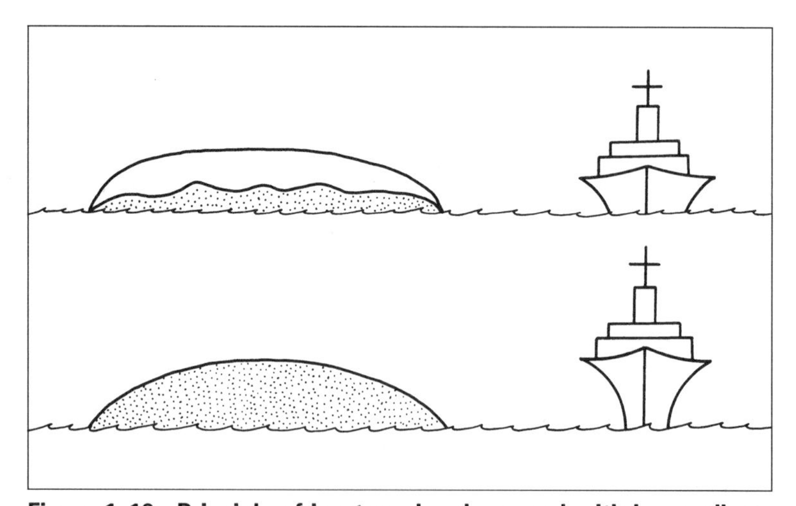

Figure 1–16 Principle of isostasy. Land covered with ice readjusts to the added weight like a loaded freighter. When the ice melts, the land is buoyed upward as the weight lessens.

Figure 1–17 Exposure of Pleistocene glacial deposits in a stream bank along the Sturgeon River, Houghton County, Michigan. Photo by W. F. Cannon, courtesy of USGS

The ice sheets appear to have taken much longer to reach their maximum extent than to recede to a comparably insignificant amount of ice as in the polar regions today. In only a geologic moment, the ice sheets collapsed and rapidly disappeared as the global climate began to warm. The most recent glacial period is the best studied of all ice ages, because of the availability of evidence. In many areas of the northern latitudes, the ice stripped off entire layers of sediment, leaving behind bare bedrock. In other areas, thick deposits of glacial till buried older deposits (Fig. 1–17), when the glaciers melted and retreated back to the poles.

The loss of greenhouse gases, principally carbon dioxide, by photosynthesis when single-celled plants first developed might have cooled the climate sufficiently to produce the first known ice age in geologic history about 2.4 billion years ago. The burial of large amounts of carbon in the Earth's crust might have been the key to the onset of perhaps the greatest of all ice ages during the late Precambrian era about 680 million years ago. The glaciations of the late Ordovician period around 440 million years ago, the middle Carboniferous period around 330 million years ago, and the Permo-Carboniferous period around 290 million years ago, might have been influenced by a reduction of atmospheric carbon dioxide to about one-quarter of its present value.

Another glacial episode that occurred about 270 million years ago might have been triggered by the spread of forests across the land as plants

TABLE 1–3 THE MAJOR ICE AGES

Time in Years	Event
10,000–present	Present interglacial
15,000–10,000	Melting of ice sheets
20,000–18,000	Last glacial maximum
100,000	Most recent glacial episode
1 million	First major interglacial
3 million	First glacial episode in Northern Hemisphere
4 million	Ice covers Greenland and the Arctic Ocean
15 million	Second major glacial episode in Antarctica
30 million	First major glacial episode in Antarctica
65 million	Climate deteriorates; poles become much colder
250–65 million	Interval of warm and relatively uniform climate
250 million	The great Permian ice age
700 million	The great Precambrian ice age
2.4 billion	First major ice age

adapted to living and reproducing out of the sea. The Earth began to cool as the forests removed atmospheric carbon dioxide, converting the carbon into organic matter that became coal, which buried substantial amounts of carbon in the crust.

Atmospheric scientists have acquired information on global geochemical cycles to understand what might have caused such a change in carbon dioxide concentration in the atmosphere. Data taken from deep-sea cores established that carbon dioxide variations preceded changes in the extent of the more recent glaciations, and possibly the earlier glacial epochs were similarly affected. The variations of carbon dioxide levels might not be the sole cause of glaciation. However, when combined with other processes, such as variations in the Earth's orbital motions, they could be a powerful influence, which might explain why the ice ages have turned on and off again throughout geologic history.

2

HISTORICAL ICE AGES

Ice ages have been infrequent, exceptional, and only occasionally widespread events in Earth's history. The first glaciation took place about 2.4 billion years ago, at a time when the planet was converting from an atmosphere rich in carbon dioxide with a strong greenhouse effect to one that contained substantial amounts of oxygen generated by marine plant life. The loss of carbon dioxide resulted in dramatically lowered global temperatures, allowing ice to creep across the globe.

The greatest ice age occurred near the end of the Precambrian era about 680 million years ago, when nearly half the landmass was shrouded in ice. Widespread glaciation also took place during the Paleozoic era, but no significant amounts of ice covered the planet during the warm Mesozoic era. The Cenozoic era was markedly cooler, prompting several glacial events. In the Pleistocene, ice ages came and went on recurring cycles of about 100,000 years. During the last glaciation, beginning about 115,000 years ago, massive ice sheets spanned the northern reaches, and ice capped virtually every mountain.

PRECAMBRIAN GLACIATION

In the early Precambrian, the sun's output was about a third less and atmospheric carbon dioxide levels were about a thousand times more than

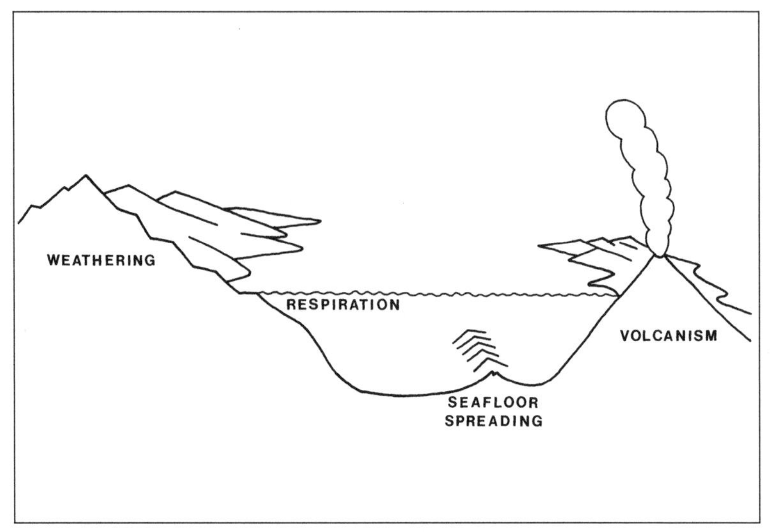

Figure 2–1 Major sources for drawing oxygen from the atmosphere and ocean.

they are now. The heavy, carbon dioxide-rich atmosphere kept the Earth quite balmy. The planet also might have retained its warmth by spinning on its axis almost twice as fast as it does today; a lack of continents, which tend to block the distribution of heat around the globe, might have also helped.

The Earth was 30 to 45 degrees Celsius warmer prior to the evolution of life, which slowly drew carbon dioxide out of the atmosphere. Carbon

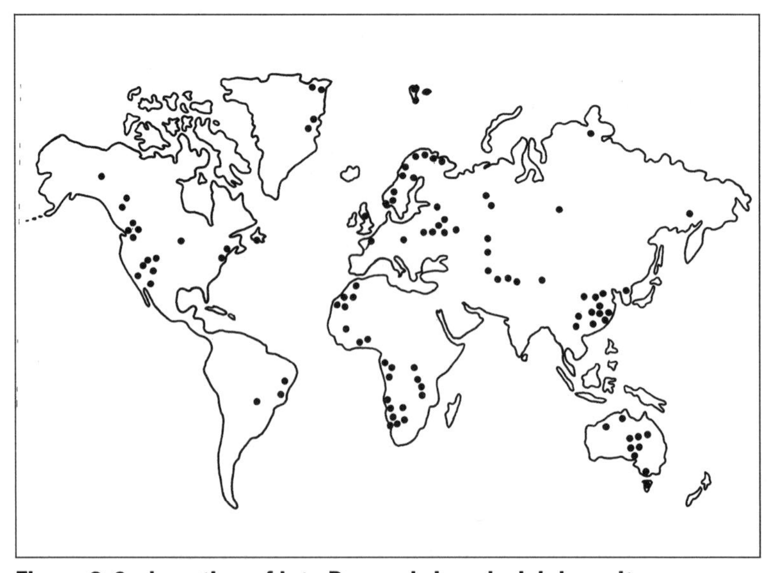

Figure 2–2 Location of late Precambrian glacial deposits.

dioxide is a greenhouse gas that prevents heat from escaping into space. This was a fortunate turn of events because without the greenhouse effect the entire planet would have acquired a thick mantle of permanent ice.

When the first microscopic plants evolved some 4 billion years ago, they began replacing carbon dioxide in the ocean and atmosphere with oxygen, so that today the relative percentages of these two gases are completely reversed. The loss of carbon dioxide dramatically cooled the global climate, and the drop in temperature initiated the first known glaciation between 2.4 and 2.3 billion years ago. Massive sheets of ice engulfed nearly the entire landmass, which comprised about 80 percent of the present continental crust.

The positions of the continents also significantly influenced the initiation of the ice ages, and landmasses drifting into the colder latitudes resulted in the development of glacial ice. Global tectonics with its extensive volcanic activity and seafloor spreading might have triggered the ice ages by drawing oxygen from the ocean and atmosphere (Fig. 2–1), preserving more organic

Figure 2–3 The supercontinent Rodinia 700 million years ago.

carbon in the sediments and preventing living organisms from returning carbon dioxide to the biosphere. Carbon burial appears to have jumped between 2.1 and 1.7 billion years ago and again between 1.1 and 0.7 billion years ago, coinciding with periods of vigorous plate tectonics.

On the ocean floor, carbonaceous sediments from the shells of once-living organisms along with the underlying oceanic crust were thrust deep into the Earth's interior at subduction zones near the edges of the continents. The growing continents stored massive quantities of carbon in thick deposits of carbonaceous rocks such as limestone. The elimination of carbon dioxide dramatically cooled the planet. Besides high rates of organic carbon burial, iron deposition and intense hydrothermal activity associated with plate tectonics lowered global temperatures. Although the middle Precambrian ice age was the first major glaciation, it was nowhere near the worst.

The burial of carbon in the Earth's crust might have prompted the onset of a second major glacial period near the end of the Precambrian, around

TABLE 2–1 ALBEDO OF VARIOUS SURFACES

Surface	Percent Reflected
Clouds, stratus	
< 500 feet thick	25–63
500–1,000 feet thick	45–75
1,000–2,000 feet thick	59–84
Average all types and thicknesses	50–55
Snow, freshly fallen	80–90
Snow, old	45–70
White sand	30–60
Light soil (or desert)	25–30
Concrete	17–27
Plowed field, moist	14–17
Crops, green	5–25
Meadows, green	5–10
Forests, green	5–10
Dark soil	5–15
Road, blacktop	5–10
Water, depending upon sun angle	5–60

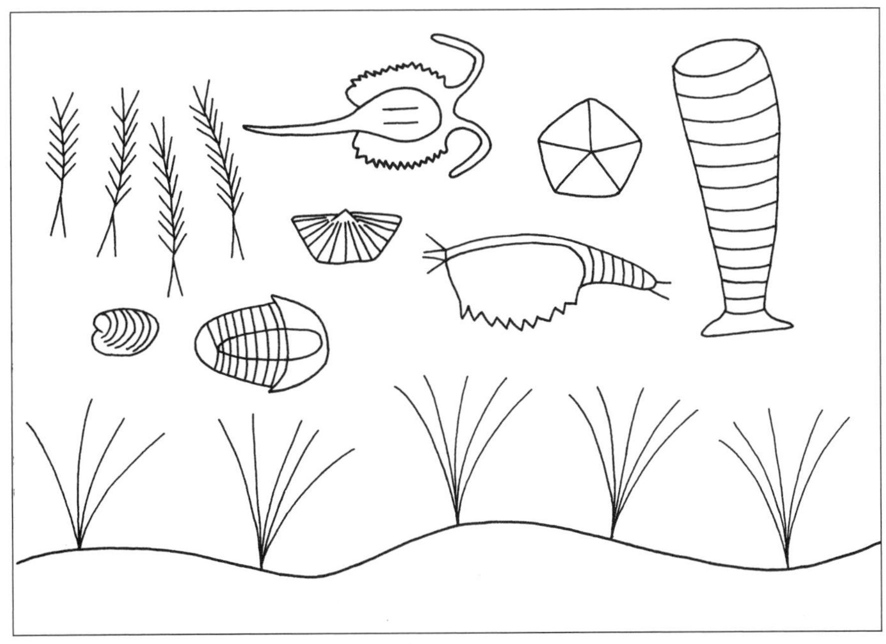

Figure 2–4 Early Cambrian marine fauna.

680 million years ago, called the Varangian ice age for the Varanger Fjord in Norway. Thick glaciers spread over the continents during perhaps the most intense glaciation the Earth has ever known, when massive ice sheets overran nearly half the land surface (Fig. 2–2). At this time, all continents assembled into a supercontinent, called Rodinia (Fig. 2–3), that probably wandered over one of the poles. Continents near the polar regions often experience extended periods of glaciation because land residing in higher latitudes has a high albedo, or reflectance, and a low heat capacity, which encourages the growth of ice.

The climate was so cold ice sheets and permafrost (permanently frozen ground) extended toward equatorial latitudes. The ice age was deadly for life in the ocean, and many simple organisms vanished during the world's first mass extinction. Animal life was still scarce, and the extinction decimated the ocean's population of acritarchs, a community of planktonic algae that were among the first organisms to develop elaborate cells with nuclei.

When the glaciation ended and the ice sheets retreated after several million years, a great diversity of animal life culminated with the evolution of entirely new species, the likes of which have never existed before or since, forever changing the composition of the Earth's biology. The rapid evolution produced three times as many phyla, which are groups of organisms sharing the same general body plan, than those living today. Life forms took off in all directions, producing many unique and bizarre creatures (Fig. 2–4).

ORDOVICIAN GLACIATION

Species that exploded onto the scene in the early Cambrian period advanced significantly in the warm Ordovician seas. The warming was due largely to the fact that the atmosphere held 16 times today's carbon dioxide content, enough to heat the climate to tropical levels even though the sun was 4 percent dimmer. The average global temperature was about 18 degrees Celsius, some 8 degrees hotter than today. Corals, which require warm waters, began building extensive carbonate reefs, and the first fish appeared in the ocean. The existence of freshwater jawless fish suggests that unicellular plants, including red and green algae, inhabited lakes and streams on the continents.

During the late Ordovician, about 450 million years ago, plants invaded the land and extended to all parts of the world. The early land plants absorbed large quantities of atmospheric carbon dioxide, and rapid burial converted the organic carbon into coal. Plants also aided the weathering process by leaching minerals from the surface rocks. Shelly marine organisms locked up substantial amounts of carbon dioxide when their shells accumulated on the ocean floor and lithified into limestone (Fig. 2–5).

The loss of carbon dioxide in the atmosphere weakened the greenhouse effect, resulting in climate cooling and extensive glaciation. The global climatic change, initiated in large part by the plant invasion, spawned a major ice age at the end of the Ordovician, about 440 million years ago. The late Ordovician glaciation apparently resulted from a drop in atmospheric carbon dioxide to about one-quarter of its present level.

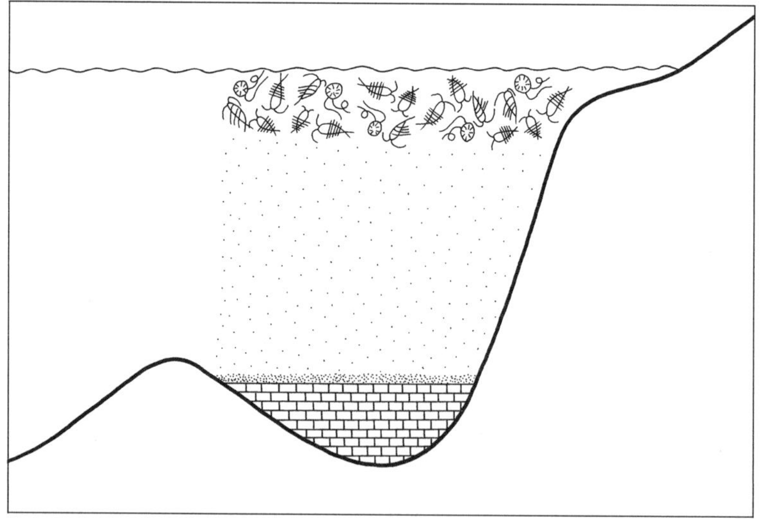

Figure 2–5 The formation of limestone from carbonaceous sediments deposited on the ocean floor.

Figure 2–6 Fossil glossopteris leaves found between layers of Carboniferous coal.

The movement of continents also shared some responsibility for the late Ordovician glaciation. Magnetic orientations in rocks from many parts of the world reveal the positions of the continents relative to the magnetic poles at various times in the geologic past. Paleomagnetic data from the African continent placed North Africa directly over the South Pole during the Ordovician.

Evidence for such widespread glaciation came from a surprising location—in the middle of the Sahara Desert. Geologists exploring for oil discovered a series of giant grooves produced by glaciers cutting into the underlying strata. Rocks embedded at the base of the glacial ice scoured the landscape as massive ice sheets moved back and forth. Other evidence that thick sheets of ice once blanketed the Sahara Desert included erratic boulders that were dumped in heaps by the glaciers along with eskers, sinuous sand deposits from glacial outwash streams.

A major mountain-building episode from the Cambrian to the Ordovician deformed areas between all continents comprising the great southern continent Gondwana, indicating their collision during this interval. Matching geologic provinces exist between South America, Africa, Antarctica, Australia, and India. The middle Paleozoic fern glossopteris, whose fossil leaves (Fig. 2–6) are found in coal beds on all southern continents, is among the best evidence for the existence of Gondwana. Matches also exist between mountains in Canada, Scotland, and Norway, indicating their assembly into the great northern continent Laurasia during this time.

The southern edge of Gondwana was in the south polar region, where an ice sheet about four-fifths the size of present-day Antarctica expanded across the continent. During the glaciation, a mass extinction eliminated about 100 families of marine animals due to the cold climate. Most victims were tropical faunas sensitive to fluctuations in the environment. Among those that became extinct were many species of trilobites. Prior to the extinction, these primitive crustaceans, a favorite among fossil collectors, accounted for about two-thirds of all species but only one-third thereafter.

At the end of the Ordovician, glaciation reached its peak, and ice sheets radiated outward from a center in North Africa. Around 430 million years ago, the ice sheets largely disappeared. As Gondwana continued drifting southward, the ice sheets became smaller. When the center of the continent neared the South Pole, the winters in the interior became colder, yet the land warmed sufficiently during the summer to melt the ice. Meanwhile, the southern glaciated edge of Gondwana moved northward into warmer seas, and the glaciers soon departed.

CARBONIFEROUS GLACIATION

Terrestrial flora first appearing some 450 million years ago were plentiful and varied during the Carboniferous period. Great coal forests of seed ferns and true trees with seeds and woody trunks (Fig. 2–7) spread across Gondwana and Laurasia in the early Carboniferous. Primitive amphibians inhabited the swampy forests, which were buzzing with hundreds of species of insects, including large cockroaches and giant dragonflies. When the climate grew colder and widespread glaciation enveloped the southern continents at the end of the period, the first reptiles emerged and began displacing the amphibians as the dominant land vertebrates.

Figure 2–7 The scale tree was one of many early trees in the lush Carboniferous forest.

Toward the end of the Carboniferous, around 290 million years ago, Gondwana hovered over the south polar regions, where glacial centers expanded across the continents, as evidenced by glacial deposits of tillites along with striations in ancient rocks. Those heavily grooved by the advancing glaciers showed lines of ice flow away from the equator and toward the poles, the opposite direction if the continents were situated where they are today. The ice would have had to flow from the sea onto the land in many areas, a highly unlikely occurrence. Instead, the southern continents drifted together over the South Pole, and massive ice sheets radiating outward from a central point crossed the present continental boundaries.

Some glacial deposits were interbedded with marine sediments. Deposits of boulders distorted the finer sediment in which they lie, indicating they fell onto the ocean's bottom muds from rafts of ice. Apparently, floating ice sheets extended outward from the land like they do today at Antarctica's huge ice shelves (Fig. 2–8). As the icebergs drifted away from the ice sheet and melted, debris embedded in the ice dropped into the sediment on the ocean floor over a wide area.

Strange, out of place boulders called erratics composed of rock types not found elsewhere on one continent, matched rocks on the opposing continent. The glacial deposits were overlain by thick sequences of terrestrial deposits, which in turn were covered by massive outpourings of basalt lava flows. Overlying these volcanic rocks were coal beds containing similar fossilized plant material.

Glacial deposits in present-day equatorial areas show that in the past these regions were much colder. Fossil coral reefs and coal deposits in the north polar regions suggest a tropical climate existed there at one time. Moreover, salt deposits in the Arctic regions indicate an ancient desert climate. Either the climate in the past altered dramatically or the continents shifted their positions with respect to the equator.

Early in the ice age, the maximum glacial effects were in South America and South Africa. As the landmass drifted southward, the chief glacial centers moved to Australia and Antarctica, indicating that the southern continents comprising Gondwana moved en masse over the South Pole. Glaciers covered large portions of east central South America, South Africa, India, Australia, and Antarctica (Fig. 2–9). In Australia, marine sediments

Figure 2–8 Unusual ice formation on the Ross Ice Shelf near McMurdo Station, Antarctica. Photo by R. F. Clayton, courtesy of U.S. Navy

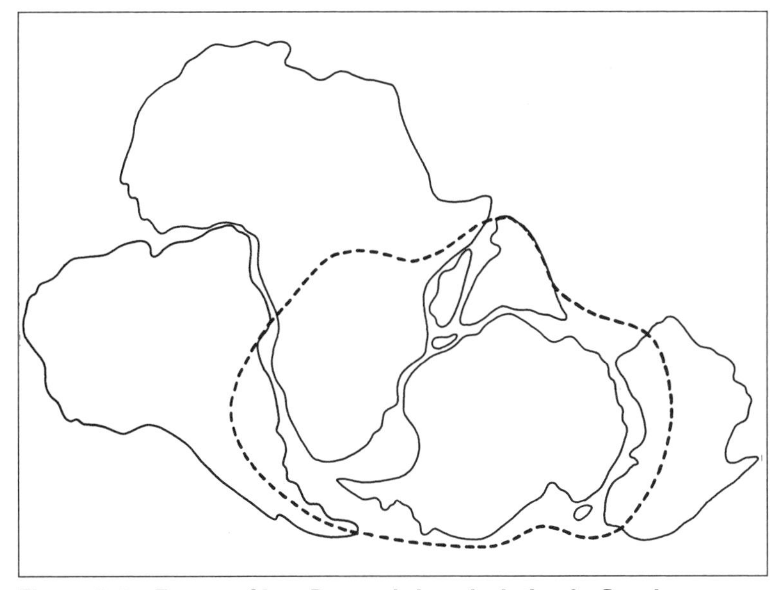

Figure 2–9 Extent of late Precambrian glaciation in Gondwana.

were interbedded with glacial deposits and tillites were separated by coal seams, indicating that periods of glaciation were interspersed with warm interglacial spells, when extensive forests grew.

Great forests spread over the land and absorbed large quantities of carbon dioxide. The lush Carboniferous forests stored the carbon in their woody tissues. Burial under layers of sediment compacted the vegetative matter and converted it into thick beds of coal, which permanently locked up carbon in the crust. The Carboniferous and Permian periods had the highest carbon burial rates of any period in geologic history. The reduction of the carbon dioxide content in the atmosphere severely weakened the greenhouse effect and cooled the climate. Land once covered with extensive coal swamps dried out as the climate continued to grow colder.

PERMIAN GLACIATION

Around 270 million years ago, the continents of Africa, South America, Antarctica, Australia, and India were heavily glaciated. In the late Permian period, all major continents combined into Pangaea (Fig. 2–10), and the collisions resulted in widespread mountain building and extensive volcanism. The interior of Pangaea was largely desert, causing the decline of the amphibians in favor of the reptiles. At the end of the Permian, the greatest extinction the Earth has ever known eliminated over 95 percent of all species. The die-off apparently occurred in two phases spaced 5 million

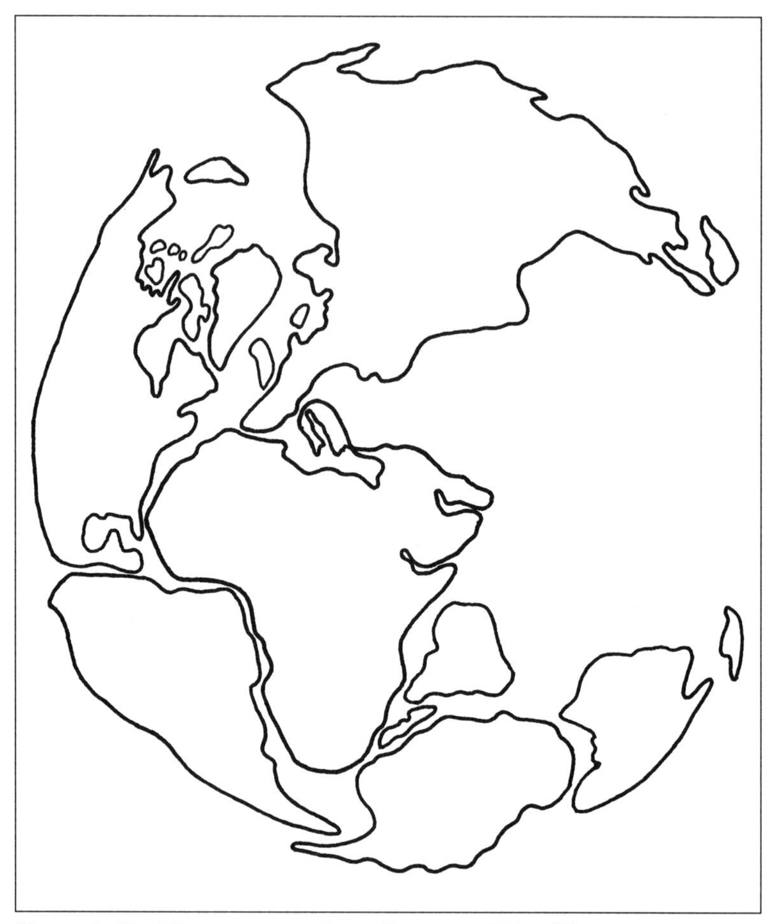

Figure 2–10 The supercontinent Pangaea 250 million years ago.

years apart. About 70 percent of the species disappeared during the first episode and 80 percent of the remaining species in the second event.

When Gondwana and Laurasia collided, creating Pangaea, the impact crumpled the crust and pushed up huge blocks into several mountain belts throughout many parts of the world. With land thrust up to high elevations, the temperature decreased and precipitation increased, spawning the growth of glaciers. Evidence for widespread glaciation include beds of Permian tillites found on nearly every continent (Fig. 2–11). Volcanoes were highly active, and unusually long periods of volcanic activity blocked out the sun with clouds of volcanic dust, significantly lowering surface temperatures.

As the continents rose higher, the ocean basins dropped lower. The changing shape of the oceans greatly influenced the course of oceanic currents, which in turn profoundly affected the climate. All known episodes of glaciation correlated with low sea levels because the massive ice sheets held vast quantities of the Earth's water. Continental margins were less extensive and narrower, confining marine habitats to near-shore areas, which might have greatly influenced the mass extinction at the end of the period. Land once covered with great coal swamps dried out as the climate grew colder, culminating in the deaths of many species.

At the beginning of the Mesozoic era, the Earth was recovering from a major ice age and the word's worst extinction event. During this time, some 450 new families of plants and animals came into existence, like a rebirth of life. Several major groups of terrestrial vertebrates made their debut, including the ancestors of modern reptiles, dinosaurs, and mammals, along with the predecessors of birds. The oldest dinosaurs originated on the southern continent Gondwana when the last of the glaciers from the great Permian ice age were departing. The region still experienced cold climates, suggesting these dinosaurs might have been warm-blooded.

Ocean temperatures also remained cool following the late Permian ice age. Episodes of climatic cooling are detrimental to species that do not adapt to the new, colder conditions or migrate to warmer regions. Marine invertebrates that escaped extinction lived in a narrow margin near the equator. Corals, which only live in warm, shallow water, were particularly hard hit as evidenced by the lack of coral reefs in the early Mesozoic. When the great glaciers melted, and the seas began to warm, reef building intensified, forming thick deposits of limestone laid down by prolific lime-secreting organisms.

Figure 2–11 Permian tillite in Roxbury conglomerate, Hyde Park, Suffolk County, Massachusetts. Photo by W. C. Alden, courtesy of USGS

TERTIARY GLACIATION

During the Tertiary period, beginning 65 million years ago, extremes in climate and topography created a greater variety of living conditions than during any other equivalent span of geologic time. The rigorous environments presented many challenging opportunities for plants and animals, and the extent to which species invaded diverse habitats was truly remarkable. It was also a time of constant change, when species had to adapt to a wide range of living conditions. The colder climate resulting from a reduction of atmospheric carbon dioxide initiated the evolution of savanna grasses and prolific grass-eating animals to feed on them.

The changing climate patterns were also influenced by the movement of continents toward their present locations, and intense tectonic activity built a variety of landforms (Fig. 2–12) and raised most mountain ranges of the world. Except for a few land bridges exposed from time to time, plants and animals were restricted from migrating from one continent to another and evolved along independent lines.

Greenland, Canada, and western Europe were one large landmass that rifted apart beginning about 85 million years ago. Greenland separated from North America and Eurasia about 57 million years ago. With the formation of Baffin Bay and the Labrador Sea, cold water from the Arctic Ocean flowed into the North Atlantic, and the change in ocean circulation dramatically altered the climate.

A narrow, curved spit of land temporarily connected South America with Antarctica, suggesting that land bridges spanned the continents as late as

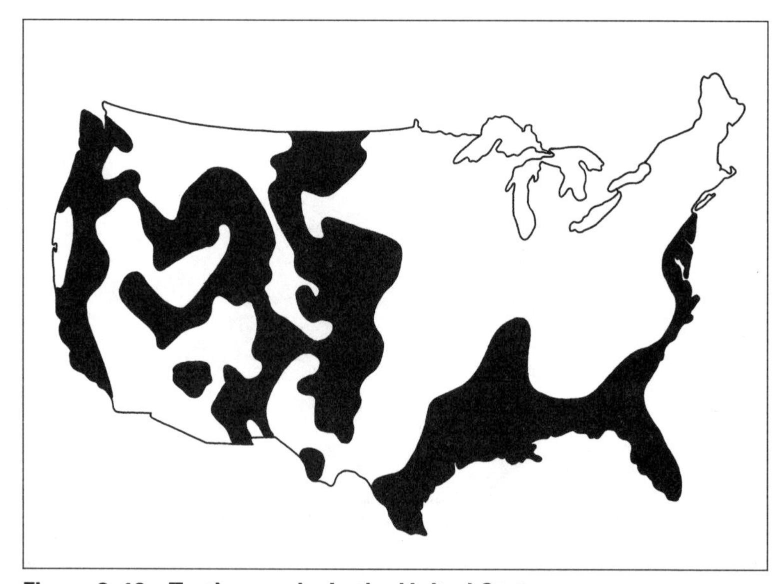

Figure 2–12 Tertiary rocks in the United States.

Figure 2–13 Wright Dry Valley, Talor Glacier region, Victoria Land, Antarctica.
Photo by W. B. Hamilton, courtesy of USGS

40 million years ago, when the sea level dropped to a near-record low, exposing land that was once beneath the ocean. Antarctica and Australia then broke away from South America and together drifted eastward. The two continents eventually rifted apart, and Antarctica moved toward the south polar region, while Australia continued in a northeastward direction. When Antarctica drifted over the South Pole, it acquired a permanent ice sheet that buried most of its terrain features (Fig. 2–13).

The Cenozoic is known for its intense mountain building, and a spurt in mountainous growth over the past 5 million years might have triggered the Pleistocene ice ages. The Rocky Mountains, extending from Mexico to Canada, heaved upward between 80 and 40 million years ago. India broke away from Gondwana early in the Cretaceous, sped across the ancestral Indian Ocean, and slammed into southern Asia about 40 million years ago, raising the Himalaya Mountains. Some 30 million years ago, Africa and Eurasia collided, initiating a major mountain-building episode that raised the Alps and other European ranges. In South America, the mountainous spine of the Andes, running along the western edge of the continent, continued to rise throughout the Cenozoic.

About 4 million years ago, Greenland, which had been largely ice free, was buried under a massive sheet of ice up to 2 miles thick. Alaska connected with east Siberia and closed off the Arctic Basin from warm water currents originating in the tropics, resulting in the formation of pack ice in the Arctic Ocean. During glacial periods, the Arctic Ocean was capped by an ice shelf hundreds of feet thick that bridged the northern continental ice sheets.

The Panama Isthmus, separating North and South America, uplifted as oceanic plates collided, spawning a lively migration of species between the two continents. Prior to the continental collision, South America was a lone island continent for 80 million years. A barrier created by the land bridge connecting the continents isolated Atlantic and Pacific species, and extinctions impoverished the once rich fauna of the western Atlantic.

The new landform halted the flow of cold water currents from the Atlantic into the Pacific, which along with the closing of the Arctic Ocean from warm Pacific currents might have initiated the Pleistocene glaciation. Never before has permanent ice capped both poles, suggesting that the planet has been steadily cooling throughout the Cenozoic.

PLEISTOCENE GLACIATION

From about 250 million to 40 million years ago, no major ice caps covered the globe, suggesting that the Pleistocene with its huge glaciers at both poles was a unique event in Earth's history. After the continents drifted to their present positions and the land rose to higher elevations, geographic conditions were ripe for a colder climate. About 3 million years ago, huge volcanic eruptions in the northern Pacific darkened the skies and global temperatures plummeted, culminating in a series of glacial episodes.

The Pleistocene epoch witnessed a progression of ice ages advancing and retreating almost as though they were cyclical. Variations in the Earth's orbital motions might have triggered the growth of continental glaciers, which partly explains the recurrence of the ice ages about every 100,000 years. Once in place, the glaciers became self-sustaining by controlling the climate. Then, mysteriously, in just a few thousand years the great ice sheets collapsed and rapidly retreated back to the poles. Analysis of deep-sea sediments and glacial ice cores (Fig. 2–14) provides a historical record of these ice ages.

The chemical composition of seawater has undergone significant variations during the Pleistocene glaciation. Between about 15 and 4 million years ago, mats of diatoms, a species of algae with shells made of silica (Fig. 2–15), spread across vast areas of the eastern tropical Pacific. The mats formed when the long, thin shells tangled together in a mesh that blanketed

Figure 2–14 Ice cores from the Ross Ice Shelf, Antarctica. Photo by Dearing, courtesy of U.S. Navy

the ocean surface in calm water. When the mats sank, they were preserved in the bottom ooze that slowly accumulated over millions of years. Because evidence of the mats does not appear in Pacific sediments younger than 4 million years old, something in the ocean must have changed significantly, causing the diatoms to disappear.

About 3 million years ago, the surface waters of the ocean cooled dramatically, causing diatoms to decline sharply in the Antarctic, presumably when sea ice reached its maximum northern extent, thus shading the algae below. Without sunlight for photosynthesis, the diatoms simply

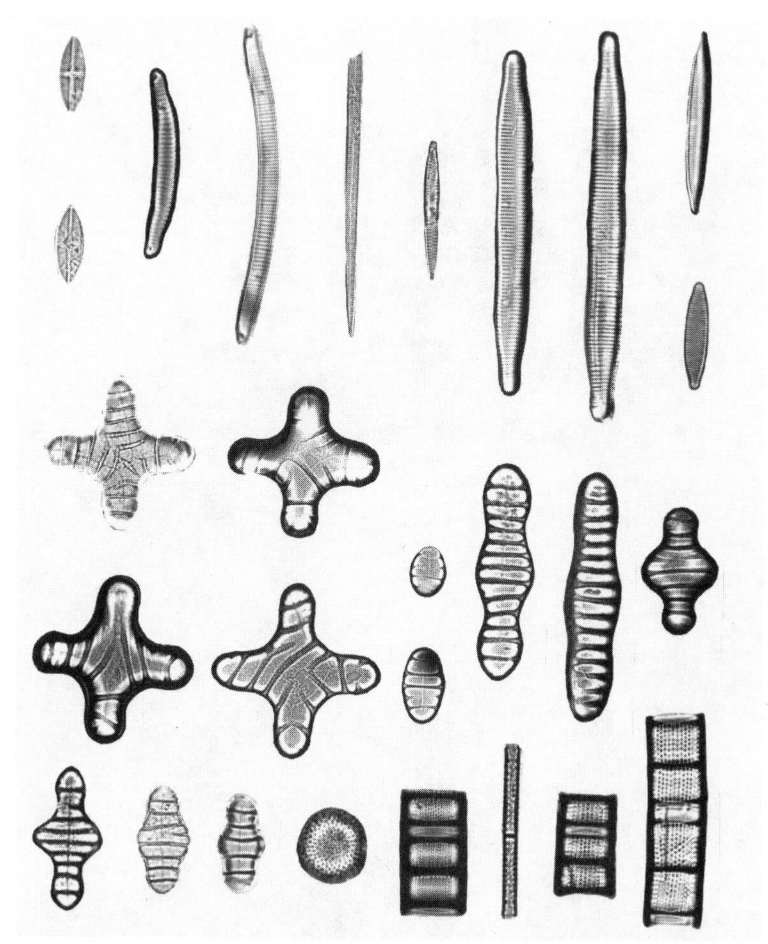

Figure 2–15 Late Miocene diatoms from the Kilgore area, Cherry County, Nebraska. Photo by G. W. Andrews, courtesy of USGS

vanished, and their disappearance marks the initiation of the Pleistocene glacial epoch in the Northern Hemisphere. Animal life on the planet also changed strikingly, and changing habitats might have spurred the evolution of early humans.

In the last ice age, massive ice sheets swept out of the polar regions, and glaciers up to 2 miles or more thick enveloped Canada, Greenland, and northern Eurasia. Overall, the glaciers covered some 11 million square miles of land that is presently ice free. The glaciation began with a rapid buildup of glacial ice some 115,000 years ago. It intensified about 75,000 years ago, possibly due to the massive Mount Toba eruption in Indonesia, and peaked about 18,000 years ago, around the time a large meteorite slammed into the Arizona desert (Fig. 2–16). During the height of the ice age, levels of atmospheric carbon dioxide were about 40 percent and methane about 50 percent less than they are today.

The cold weather and approaching ice forced species to migrate to warmer latitudes. Ahead of the advancing ice sheets, which traveled upward of several hundred feet a year, lush deciduous woodlands gave way to evergreen forests, which yielded to grasslands that later became barren tundra and rugged periglacial regions on the margins of the ice sheets. Approximately 5 percent of the planet's water was locked up in glacial ice. The accumulated ice dropped sea levels about 400 feet and shorelines advanced seaward up to 100 miles or more. The drop in sea level exposed land bridges and linked continents, spurring a vigorous migration of species, including humans, to various parts of the world. Indeed, we are products of the ice ages.

The continental ice sheets contained approximately 10 million cubic miles of water and covered about one-third the land surface with glacial ice, three times its current size. Land shrouded by glaciers sank hundreds of feet due to the great weight of the ice sheets, which caused the crust to sag. The added pressure squeezed the upper mantle, forcing it aside. As the deep rock of the mantle flowed from the ice-filled depression, it raised a bulge about 300 feet high beyond the margin of the ice sheet.

When most of the ice melted between 14,000 and 7,000 years ago, the Earth began rebounding to its preglacial shape. Even today, the northern

Figure 2–16 Meteor Crater, Coconino County, Arizona. Photo by W. B. Hamilton, courtesy of USGS

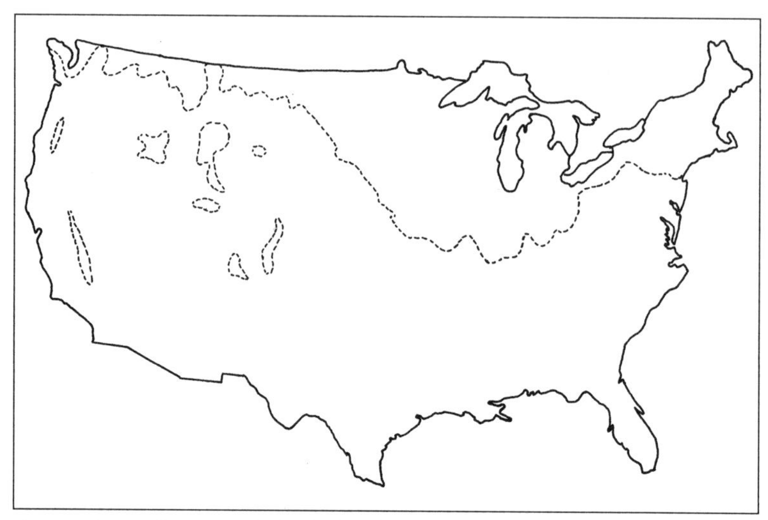

Figure 2–17 Extent of glaciation in the United States during the last ice age.

lands are rising as much as half an inch a year, long after the massive weight of the glaciers had been lifted. A record of this rebound exists on coastlines around the world, where changing sea levels have cut a series of terraces.

During the last ice age, North America held two main glacial centers. The largest glacier, called the Laurentide, the biggest and thickest of the ice sheets ringing the North Atlantic, blanketed an area of about 5 million square miles. It extended from Hudson Bay and reached northward into the Arctic Ocean and southward into eastern Canada, New England, and the upper midwestern United States (Fig. 2–17). Several times during the glaciation, the Laurentide ice sheet partially collapsed, sending great armadas of icebergs into the North Atlantic.

A smaller ice sheet, known as the Cordilleran, originated in the Canadian Rockies and engulfed western Canada and the northern and southern sections of Alaska, leaving an ice-free corridor down the center of the state used by humans migrating from Asia into North America. Scattered glaciers also covered the mountainous regions of the northwestern United States. Ice buried the mountains of Wyoming, Colorado, and California, and rivers of ice linked the North American cordillera with mountains in Mexico.

Two major ice sheets engulfed much of Europe as well. The largest, called the Fennoscandian, fanned out from northern Scandinavia and covered most of Great Britain as far south as London and large parts of northern Germany, Poland, and European Russia. A smaller ice sheet, known as the Alpine and centered in the Swiss Alps, enveloped parts of Austria, Italy, France, and southern Germany. In Asia, glaciers occupied the Himalayas and blanketed parts of Siberia.

In the Southern Hemisphere, only Antarctica possessed a major ice sheet, which was about 10 percent larger than it is today, extending as far as the tip of South America. Sea ice surrounding Antarctica expanded to nearly double its modern wintertime area. Smaller glaciers capped the mountains of Australia, New Zealand, and the Andes of South America, which contained the largest of the southern alpine ice sheets. Throughout the rest of the world, mountain glaciers topped peaks that are currently ice free.

At both poles, the excess ice had nowhere to go except into the sea, where it calved off to form icebergs (Fig. 2–18). During the peak of the last ice age, icebergs covered half the area of the ocean. The ice floating in the sea cooled the ocean even in the tropics, which dropped 2 degrees or more Celsius. The floating ice also reflected sunlight back into space, thereby maintaining a colder climate with average global temperatures about 3 to 5 degrees lower than today, perhaps colder than at any time in the past 65 million years.

As many as nine periods of abrupt warming punctuated the ice age. These periods are called Dansgaard-Oeschger events after their discoverers, Danish geologist Willi Dansgaard and the Swiss geologist Hans Oeschger. Ice core data taken from the Greenland ice cap have shown that throughout the ice age the climate apparently warmed abruptly and then gradually cooled back to full ice age conditions. During intervals of 6,000 to 10,000 years apart, the continental ice sheets suddenly collapsed, sending vast arrays of icebergs crashing into the sea. These iceberg invasions, known as Heinrich events for Hartmut Heinrich, the German geologist who discovered them, occurred at the end of each series of increasingly intense warmings, called interstadials. Five major Heinrich events occurred between 70,000 and 14,000 years ago during the coldest and longest spans of the Dansgaard-Oeschger cycles.

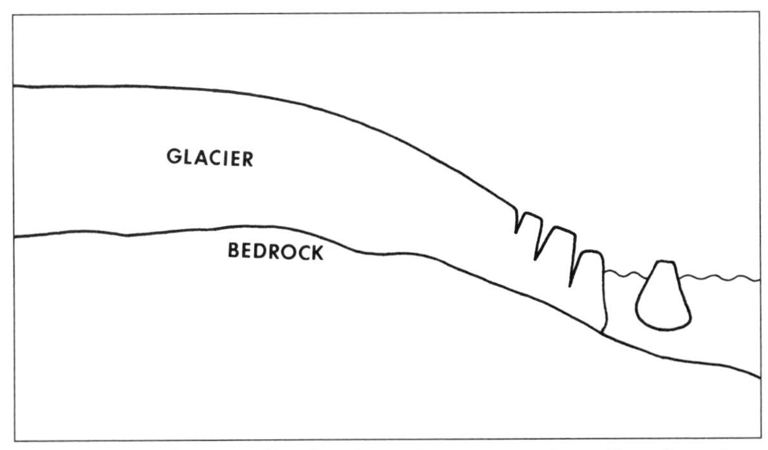

Figure 2–18 Glaciers flowing into the ocean calve off to form icebergs.

The Arctic Ocean has long been thought to have held only thin sea ice during the ice age. But the discovery on the Arctic seafloor of long grooves carved by ancient, giant icebergs much larger than those of today raises the possibility that a floating ice shelf up to 2,000 feet thick covered much of the Arctic Ocean. This great expanse of ice would have connected the glacial sheets covering North America and northern Europe.

Because of the lowered temperatures, less water evaporated from the sea, causing precipitation rates over the land to fall dramatically. Since little melting occurred during the cooler summers, less snowfall was required to sustain the ice sheets. The lower precipitation levels also increased the spread of deserts across many parts of the world. Desert winds blew much more intensely than they do today, producing gigantic dust storms. So much dust clogged the atmosphere, it blocked out sunlight and significantly shaded the Earth. The dust falling into the oceans fertilized it with iron, spawning prolific blooms of floating plant life called phytoplankton, which drew carbon dioxide out of the atmosphere, thereby weakening the greenhouse effect.

These processes provided an efficient feedback mechanism for the continuation of the glacial ice. Considering the high solar reflectance of the ice sheets themselves, which also tended to cool the planet, how the ice age ended remains a great mystery.

3

THE INTERGLACIAL

Each ice age was followed by a relatively short, warm interglacial period (Fig. 3–1), an abnormal warming spell between glaciations, a respite before the next glacial assault. Indeed, ice ages appear to be more the norm than the exception over the last 3 million years.

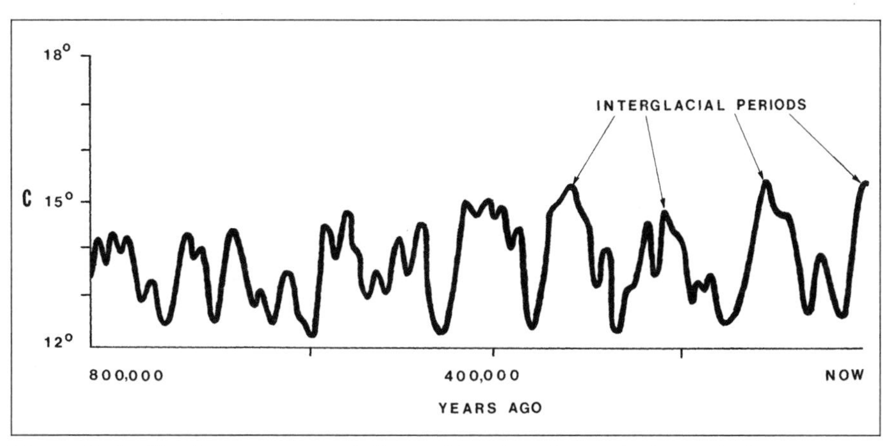

Figure 3–1 The ice ages were followed by short interglacial periods.

One of the most dramatic climate changes in geologic history occurred during the present interglacial, known as the Holocene epoch, beginning about 10,000 years ago. The cold conditions of the ice age gave way today's warmer climate. As the glaciers melted and retreated back to the poles, massive floods of meltwater flowed into the ocean, raising sea levels several hundred feet. This period is also commensurate with the rise of civilization, due in large part to its optimal climatic conditions, which allowed the development of agriculture.

THE HOLOCENE EPOCH

The rapid melting of the great ice sheets left many unsolved questions, and the changing of habitats during deglaciation also caused numerous species of large mammals to mysteriously disappear. One such question was prompted by the discovery of hippopotamus and crocodile bones in the middle of African deserts. When the ice sheets began to shrink at the end of the last ice age, Africa and Arabia, which were tropical regions at that time, eventually dried out. The climatic change caused by the rapid warming culminated with the expansion of the arid regions between about 14,000 and 12,500 years ago.

During a wet period from about 12,000 to 6,000 years ago, many of today's African deserts supported lush vegetation and contained several large lakes, where today only sand lies. Lake Chad, on the southern border of the Sahara Desert, was Africa's largest lake, covering an area many times its present size. Other lakes around the world were similarly affected. African swamps, long since vanished, once harbored large populations of hippopotamuses and crocodiles, whose bones now bake in the hot desert sun.

Inland seas filled with sediments, and subsequent uplifting drove out the waters, leaving behind salt lakes. Utah's Great Salt Lake is only a remnant of a vast sea that covered the interior of North America during the Mesozoic era. Sediments eroded from the western highlands flowed into the sea, providing the American West with much of its rugged beauty (Fig. 3–2).

During a rainy period between 12,000 and 6,000 years ago, the lake expanded several times its current size and flooded nearby salt flats. The long wet spell following the retreat of the glaciers might have been due to intense monsoons that brought moisture-laden sea breezes inland over Africa, India, and Southeast Asia. The interior of the continents 9,000 years ago warmed more in the summer, which strengthened the monsoon winds and increased rainfall.

In North America, the glaciers radically changed atmospheric circulation and rerouted storm patterns. High-pressure centers over the Laurentide ice sheet produced strong easterly winds along the southern flank of the glacier. The climatic change provided the Mojave and nearby deserts of the southwestern United States adequate rainfall for woodlands to grow after

Figure 3–2 Big Horn Basin near Lovell, Big Horn County, Wyoming. Photo by J. R. Balsley, courtesy of USGS

the retreat of the ice sheets. Similar wet spells occurred in African and Indian deserts.

Between 9,000 and 6,000 years ago, when glaciers over North America were shrinking, precipitation over much of the midwestern United States dropped as much as 25 percent, while the mean July temperature rose as much as 2 degrees Celsius. Apparently, the postglacial eastern and southeastern United States was not much warmer 6,000 years ago than those areas are today, probably due to the moderating effect of the Atlantic Ocean.

The return of the warm ocean currents prompted a second episode of melting that led to the present volume of ice by about 6,000 years ago, when the glaciers reached their farthest northern retreat. The climate change resulted in a period of unusually warm, wet conditions called the Climatic Optimum between 6,000 and 4,000 years ago, when many regions of the world warmed on average about 3 degrees. The melting ice caps released a torrent of floodwater into the sea and raised sea levels 300 feet higher than when the Holocene began. Following the receding ice sheets, plants and animals returned to the northern latitudes.

About 4,000 years ago, global temperatures dropped significantly and the climate became drier, forming today's deserts (Fig. 3–3). About 2,500 years ago, the climate in Antarctica cooled halfway to ice age levels and did not

Figure 3–3 Saguaro and other desert vegetation on the west slope of Superstition Mountains, Arizona. Photo by B. Brixner, courtesy of USDA–Soil Conservation Service

TABLE 3–1 MAJOR DESERTS

Desert	Location	Type	Area (square miles × 1000)
Sahara	North Africa	Tropical	3,500
Australian	Western/interior	Tropical	1,300
Arabian	Arabian Peninsula	Tropical	1,000
Turkestan	Central Asia	Continental	750
North America	Southwestern U.S./ Northern Mexico	Continental	500
Patagonian	Argentina	Continental	260
Thar	India/Pakistan	Tropical	230
Kalahari	South-West Africa	Littoral	220
Gobi	Mongolia/China	Continental	200
Takla Makan	Sinkiang, China	Continental	200
Iranian	Iran/Afganistan	Tropical	150
Atacama	Peru/Chile	Littoral	140

recover for another 700 years. Around 1,000 years ago, the world warmed again, during a period called the Medieval Climate Maximum. Some 500 years later, the Earth plunged into an extended cold spell called the Little Ice Age, when average global temperatures dropped about 1 degree. Other Little Ice Age events occurred during the Holocene on a 1,500-to-3,000-year cycle. The climate also changed on time scales between 1,000 and 15,000 years.

DEGLACIATION

After some 100,000 years of gradual accumulation of snow and ice up to 2 miles thick in the higher latitudes of North America and Eurasia, the giant glaciers suddenly melted away in a matter of only a few thousand years, retreating upward of half a mile or more annually. At least one-third of the glacial ice melted between 14,000 and 12,500 years ago, when average global temperatures increased about 5 degrees to nearly present-day levels.

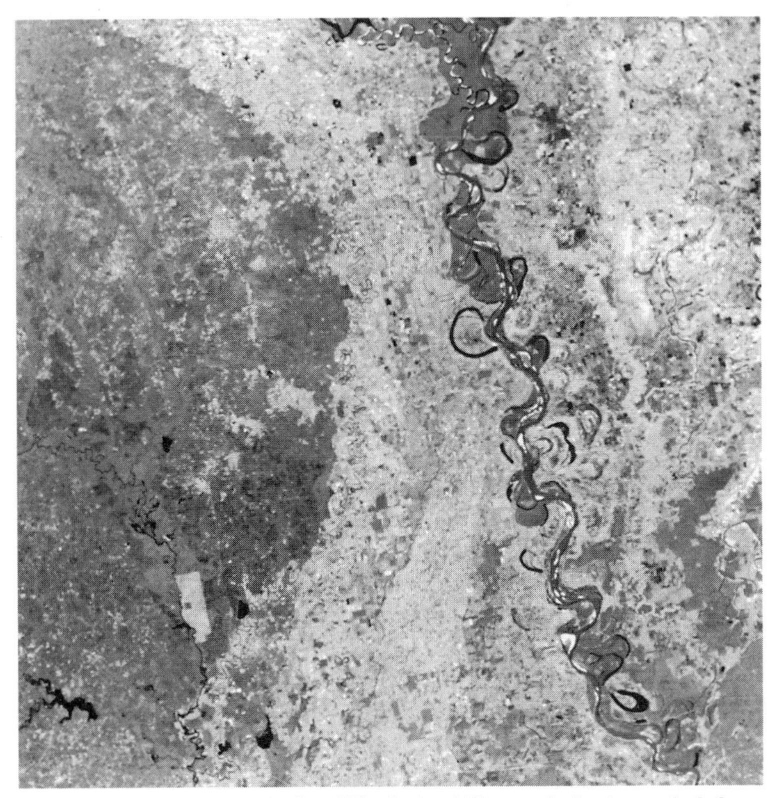

Figure 3–4 The meandering Mississippi River winds through Arkansas, Mississippi, and Louisiana. Courtesy of NASA

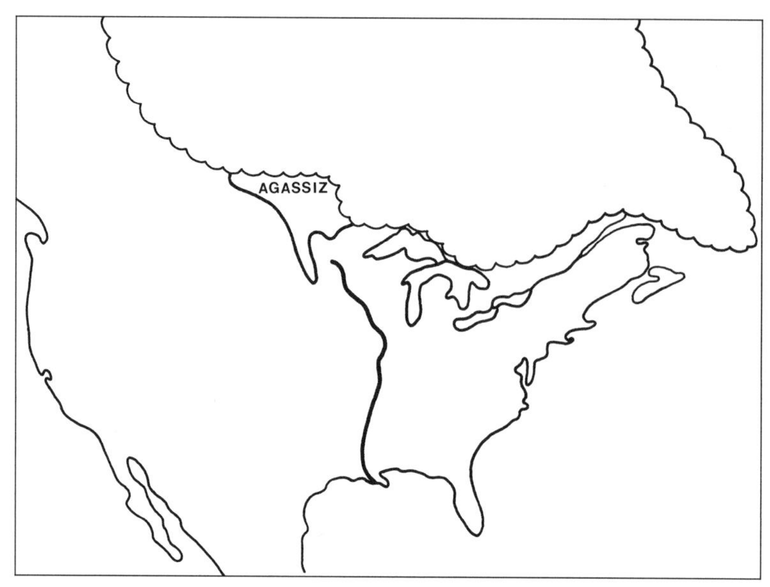

Figure 3–5 Location of Lake Agassiz during the last ice age.

The melting of the great ice sheets sent massive floods of meltwater raging across the land, as water gushed from trapped reservoirs beneath the glaciers. When the North American ice sheet retreated northward, most of the meltwater laden with sediment from its southern edge flowed down the Mississippi River toward the Gulf of Mexico, widening the channel several times its present size (Fig. 3–4). Other rivers overreached their banks to carve out new floodplains and lay down new soil.

Huge lakes of meltwater trapped below the ice sheet broke loose and rushed down river valleys in huge torrents to the Gulf of Mexico and the Atlantic Ocean, surpassing the flow of the Amazon River of South America, the largest in the world. The floods continued until the weight of the ice sheet shut off the outlet of the reservoir. When the water pressures built up again, another massive surge of meltwater spouted from beneath the glacier, rushing toward the sea in giant waves. While flowing under the ice, the water surged in vast turbulent sheets that scoured deep grooves in the surface, forming steep ridges carved out of solid bedrock. Several times, huge volumes of meltwater escaped from under the ice to further sculpt the landscape.

A vast reservoir of meltwater, larger than any of the existing Great Lakes, called Lake Agassiz (Fig. 3–5) for Louis Agassiz the foremost ice age proponent, formed in a huge bedrock depression carved out by the great ice sheets in today's southern Manitoba. Prior to about 11,000 years ago, the lake overflowed a rise in the bedrock that formed a dam and drained down the Mississippi River.

After the ice sheet retreated beyond the Great Lakes, which were themselves carved out by the glaciers, a channel opened to the east, and the water flowed across the Great Lakes region and down the St. Lawrence River, flooding the Gulf of Maine. The diversion shut off most of the meltwater discharging down the Mississippi to the Gulf of Mexico. During this time, the Niagara River Falls began cutting its gorge (Fig. 3–6) and has traversed more than 5 miles northward since the end of the last ice age.

The cold freshwater entering the North Atlantic rerouted ocean circulation, initiating a return to ice age conditions and a pause in melting known as the Younger Dryas, named for an Arctic wildflower that colonized postglacial Europe. The unusual cold spell struck northern Europe, Greenland, and the Atlantic coast of Canada from about 11,500 to 10,700 years ago, sending these regions back into near glacial conditions. The system of ocean currents that warms Europe ceased operating, and sea ice covered much of the North Atlantic during this period.

Figure 3–6 Niagara Falls was created as the Niagara River flowed from Lake Erie northward into Lake Ontario. Courtesy of NASA

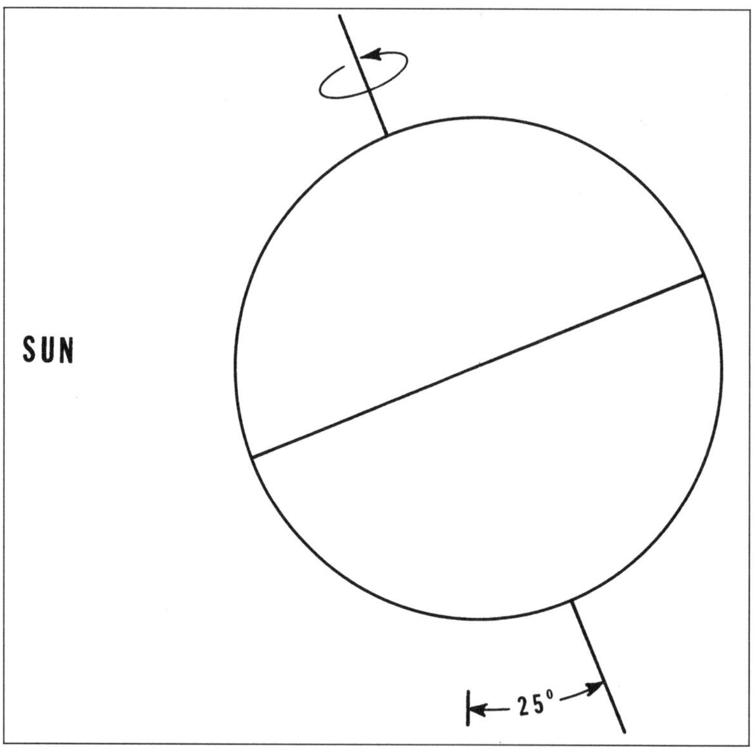

Figure 3–7 About 9,000 years ago, the Earth's axis tilted more steeply toward the sun, causing a greater difference between seasons.

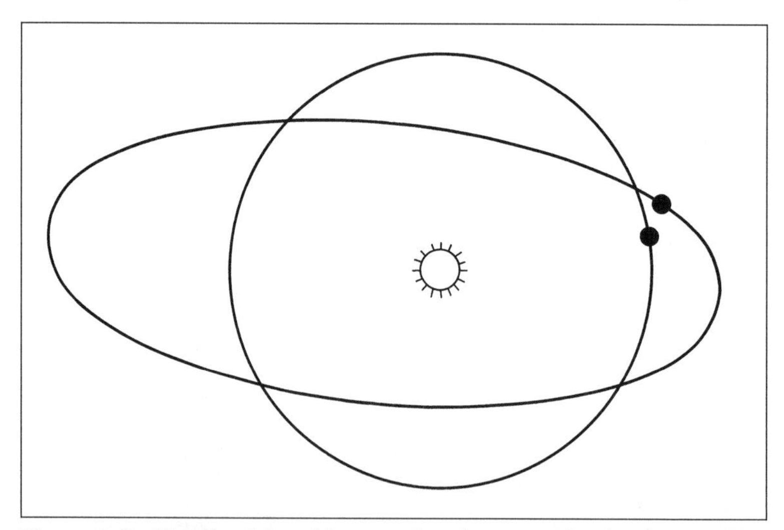

Figure 3–8 The Earth's orbit stretches from nearly circular to elliptical and back again in about 100,000 years, making the distance from the sun vary by about 11 million miles.

The rapid melting of the glaciers culminated in the extinction of single-celled marine animals called foraminifera, which met their demise when a torrent of meltwater and icebergs spilled into the North Atlantic. The massive floods formed a cold freshwater lid on the ocean that substantially changed the salinity of seawater, whose lower density kept it from sinking into the abyss. The cold waters also blocked poleward flowing warm currents from the tropics, causing land temperatures to fall to near ice age levels.

The flow of cold water into the North Atlantic shut off the conveyor belt of warm water from the tropics, spurring a return to the ice age conditions of the Younger Dryas. About 1,000 years later, a lobe of ice advanced across the western end of the Lake Superior basin and blocked the exit of meltwater to the east. As a result, Lake Agassiz rose again, diverting its meltwater back down the Mississippi, reactivating the flow of warm ocean currents, which spawned a second bout of melting.

Massive lakes formed by the retreating glaciers also might have catastrophically drained into the North Atlantic, disrupting deep-water formation and the transfer of heat northward. A renewal of the deep-ocean circulation system, which shut off entirely or became weakened during the ice age, might have thawed the planet out of its deep freeze. After the ice age's final cold spell, temperatures rapidly climbed to present interglacial conditions within less than a century.

The rapid deglaciation might have been driven largely by forces other than simply warming of the climate. The ultimate cause for bringing the last ice age to an abrupt end might have been changes in the Earth's orbit and tilt of its rotational axis. About 9,000 years ago, the planet's axis tilted more steeply toward the sun than today (Fig. 3–7). The Earth came closest to the sun in the late July instead of early January as it does now, and the seasonal range of distances between the sun and Earth was greater (Fig. 3–8).

In the Northern Hemisphere, which contained most of the glacial ice, the Earth received 7 percent more solar radiation in July, making summers warmer than at present. Furthermore, the high latitudes received the maximum amount of insolation (input of solar radiation) because the rotational axis was more inclined toward the sun. The net effect was a 15 percent difference in the amount of sunlight received over much of the Northern Hemisphere during summer compared to that at wintertime, which was about twice the difference it is today, making the contrast between seasons much greater. Since 9,000 years ago, the seasonal climate extremes gradually decreased to more modern values.

PALEOCLIMATE INDICATORS

Many of the most baffling questions about the Holocene climate are answered by analyzing pollen grains recovered from ancient bogs and lake bed sediments as well as other paleoclimate indicators, including plant fossils found in pack rat middens (refuse piles). The study of ancient plant pollen and spores is known as the science of palynology. Pollen and spores can survive in sediments and rock for thousands and even millions of years and provide a portrait of the plant life that once covered an area.

Analysis of ancient plant and animal remains can determine the climatic conditions of the past by comparing them to present-day relationships between species distribution and climate. For example, a combination of spruce and sedge fossil pollen is an indicator of a cold, dry climate, whereas fossil pollen of leafy herbs typical of the prairie of the midwestern United States suggests a warm, dry climate.

Another important continental climate indicator is lake-level fluctuations. Lakes act as natural rain gauges, and ancient rainfall amounts are implied by studying the positions of past shorelines and depth indicators such as the mineral, floral, and faunal composition of lake sediments. Lake-level records are particularly valuable for the arid regions of Africa, Australia, and the western United States because pollen records are sparse in these regions.

Figure 3–9 A tree sample being prepared for annual growth ring studies. Photo by L. E. Jackson, Jr., courtesy of USGS

Figure 3–10 The General Sherman Tree is a giant sequoia that is one of the largest trees on Earth. Courtesy of National Park Service

An additional paleoclimate indicator is the size of sand grains deposited over the past 260,000 years in Lake Biwa, Japan, one of the oldest lakes in the world. The rise and fall of the sediment size kept pace with the advances and retreats of the great ice sheets. The grain sizes reflected past erosion rates, which in turn depended on climatic changes in rainfall, temperature, and wind speed. High precipitation rates increase erosion and consequently the amount of coarse sand grains carried into the lake. Tropical lakes in the Northern Hemisphere were similarly affected. Ancient riverbeds carrying alluvium to the sea are another source of information on changing conditions of the past. The climate data comes from records of extreme floods that left lasting marks on major rivers of the world.

The analysis of tree rings (Fig. 3–9) provides one more useful indicator of past climates because the wider the rings generally the more favorable the climate. During a drought or an unusually cool season, a tree's growth is often stunted, which is reflected by a reduction in its ring width. Tree

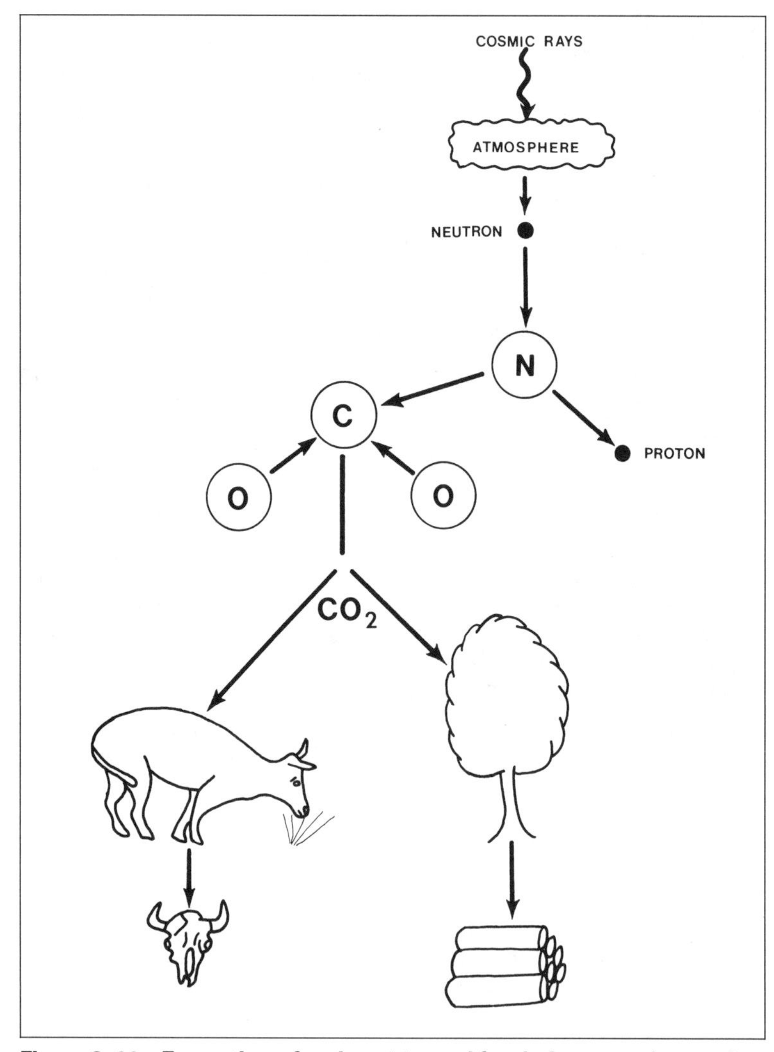

Figure 3–11 Formation of carbon 14, used for dating organic remains.

rings of ancient, well-preserved trees tell of a region's climate history. By analyzing tree rings of the bristlecone pine, among the longest living plants on Earth, a drought index for the western United States is provided dating to the year 1600. The tree ring data also can reconstruct a regional series of annual precipitation. Trees living between 1645 and 1715 gave an account of anomalous climatic conditions that coincided with the Maunder Minimum of sunspot activity during the Little Ice Age.

Tree rings samples also can be obtained by taking small cores from living trees such as the great sequoias of northern California (Fig. 3–10), which are among the world's oldest living plants. Furthermore, by analyzing the radioactive carbon 14 in rings from ancient, well-preserved trees, a history of the carbon 14 content of the atmosphere has been reconstructed dating back more than 7,000 years. A continued study of tree rings could eventually provide important data for testing solar and lunar influences on the climate.

The quantity of carbon 14 generated in the atmosphere varies directly with the amount of solar cosmic rays, which are more abundant during times of intense solar activity. The cosmic rays striking nitrogen atoms cause them to mutate into carbon 14 atoms. Plants and animals absorb the carbon 14 along with ordinary carbon 12 (Fig. 3–11), and after death the carbon 14 clock starts ticking, providing an accurate method of dating organic remains. With improved analytical techniques, carbon dating has been extended as far back as 100,000 years ago and can date events taking place during the last ice age.

Paleoclimate studies, which analyzed marine microfossils in the cores of seafloor sediments for their carbon dioxide levels, suggest that the previous warm interglacial called the Eemian, which began about 130,000 years ago and abruptly ended about 15,000 years later, might have been warmer than the present one. Apparently, higher levels of atmospheric carbon dioxide during the preceding interglacial led to greater greenhouse warming. Evidence from seafloor sediments, tree rings, pollen samples, and historical records shows that conditions have fluctuated little during the Holocene, suggesting that interglacials in general might escape the wild climatic variations of the ice ages.

Ice cores taken from Greenland's ice cap span a period of 250,000 years, in which the Earth has experienced an interglacial period, weathered an ice age, thawed during another interglacial, slipped back into another ice age, and warmed again during the present interglacial. The cores provided the first reliable ice record of the Eemian interglacial. Past ice core work has focused on the most recent ice age, when the climate fluctuated rapidly between mild and cold times.

The climate warmed quickly at the beginning of the Eemian and remained temperate for about 3,000 years. Instead of a period of stable warmth, however, the Eemian was divided into three warm stages, called

Figure 3–12 An irrigated cornfield in Hamilton County, Nebraska. Photo by G. Alexander, courtesy of USDA–Soil Conservation Service

interstadials, broken up by two long spells of extensive cold, called stadials, lasting 2,000 and 6,000 years each. The cold temperatures averaged 7 degrees Celsius lower than the warm parts of the Eemian, which was, on average, 2 degrees warmer than today's interglacial, when the climate has been remarkably stable over the last 8,000 years. The climate remained stable enough for people to develop the agriculture needed to maintain an advanced society, during perhaps the only time it could have been done (Fig. 3–12).

Apparently, major changes in the ocean's circulation triggered the cold spells of the Eemian. Periodic halts in the northward flow that carries warm Atlantic water to the far northern North Atlantic are a possible cause of the climatic oscillations around the end of the last ice age as well. The shutdowns in the ocean current might have stemmed from occasional collapses of the ice sheets into the ocean.

Geologic records indicate that ice sheets began building in the high latitudes of North America and Eurasia at a time when the climate was at least as warm as today. Mild conditions, especially during the winter, led to increased evaporation from the ocean, inducing more snowfall over the northern regions. The snow survived during the summer because of shifts in the Earth's orbit, which reduced the amount of sunlight reaching the Arctic during that season. The Earth's orbit aids in the buildup of glacial ice with slight variations in the orientation of the rotational axis and its distance from the sun. These orbital variations decrease the amount of solar radiation reaching the northern latitudes, spawning the growth of ice sheets.

Paleoclimate studies indicated that glacial periods hung around many times longer than interglacials. The previous four interglacials lasted about 8,000 to 15,000 years. The fact that the present interglacial is already some 10,000 years old suggests it probably has just about run its course and the next ice age might be looming one or two thousand years away. Perhaps the gradual global warming over the last 140 years has kept the ice age at bay.

Although the preceding interglacial appears to have been warmer than the present one, the warm climate was powerless to halt the last ice age. Even with today's industrialization, destruction of the forests, and extension of agriculture, which pumps large quantities of greenhouse gases into the atmosphere and produces global warming, human activity might not in the long run save civilization from advancing ice sheets. This is because melting ice caps are inherently unstable, and shifting weather patterns might have been important for glacial inception during the entire Pleistocene.

Once a large portion of the Northern Hemisphere is snowbound all year long, the increased albedo reflects solar radiation back out into space that normally would warm the Earth. This makes the ice sheets self-perpetuating, which explains why ice ages linger for so long. It also makes interglacials short-lived events during the last million years or so. Once an interglacial becomes established, it exists only as long as the darker, snow-free ground continues to absorb warmth from the sun, which prevents glaciers from expanding across whole continents.

POSTGLACIAL EXTINCTIONS

At the end of the last ice age, an unusual extinction event killed off large terrestrial plant-eating mammals weighing over 100 pounds called megaherbivores, many of which weighed up to a ton or more. The wave of extinctions occurred between 13,000 and 10,000 years ago, with the greatest die out peaking around 11,000 years ago. Woolly rhinoceroses, mammoths, and Irish elk disappeared in Eurasia, and the great buffalo, giant hartebeests, and giant horses disappeared in Africa. Australia lost its giant marsupials, including the giant kangaroos, and over 80 percent of its large mammals in addition to a significant number of bird species.

In North America, 35 classes of mammals and 10 classes of birds became extinct. The giant ground sloths, mastodons, and woolly mammoths disappeared, with a possible exception of the dwarf woolly mammoth, which might have survived in the Arctic until about 4,000 years ago. The loss of these animals also forced into extinction their main predators: the American lion, saber-toothed tiger, and dire wolf.

Around 11,000 years ago, many parts of North America were occupied by ice age peoples, whose spear points were found among the remains of giant mammals, including mammoths, mastodons, tapirs, native horses, and camels. These people crossed into North America from Asia (Fig. 3–13) over a land bridge formed by the draining of the Bering Sea and moved through an ice-free corridor east of the Canadian Rockies. When they entered North America, they found a land densely populated with large mammals similar to those decimated in Europe and Asia.

Instead of migrating to North America in several waves, however, small bands of nomadic hunters probably crossed the ancient land bridges in pursuit of game and ended up in the New World purely by accident. The human hunters arriving from Asia sped across the virgin continent following migrating herds of large herbivores, leaving these animals' carcasses along the way. Extinctions were also massive in Australia, possibly perpetrated by the ancestors of the Aborigines, who crossed over from Java about 40,000 years ago.

The global environment reacted to the changing climate at the end of the last ice age with declining forests and expanding grasslands. The climatic change disrupted the food chains of many large animals, and deprived of their food resources they simply vanished. Also by this time, humans had

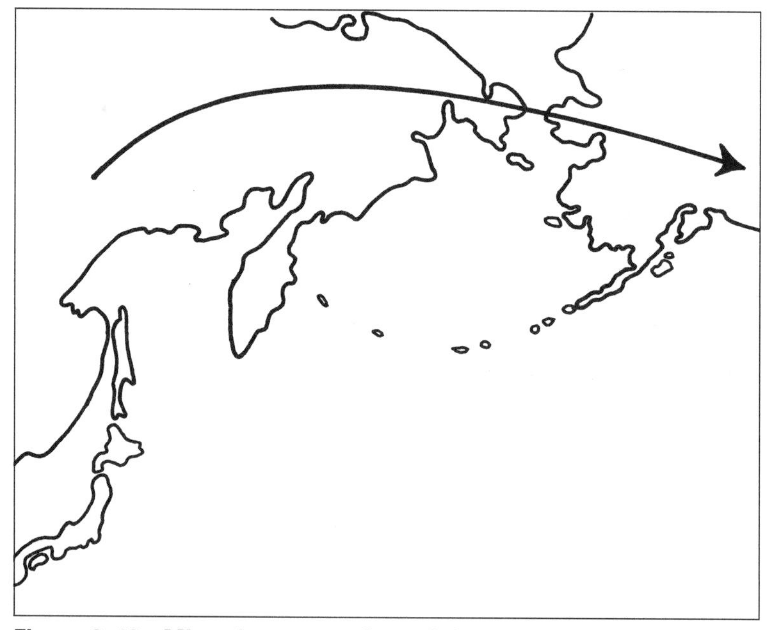

Figure 3–13 Migration routes from Asia to North America by the Bering Strait during the last ice age.

become proficient hunters and roamed northward on the heels of the retreating glaciers.

The transition from ice age conditions to the present warm interglacial period caused sea levels to rise and lake levels to fall over much of North America. Large mammals congregating at the few remaining water holes might have been vulnerable to human hunting pressures. With plentiful prey and little exposure to new diseases, human populations soared, and people spread southward, leaving behind many big-game extinctions.

Unlike earlier episodes of species extinction, such as the disappearance of the dinosaurs, this event did not significantly affect small mammals, amphibians, reptiles, and marine invertebrates. It seems strange that after enduring several previous periods of glaciation over the past 2 to 3 million years, these large mammals should suddenly fall victim following the last ice age. Perhaps the only exception were dwarf mammoths living on Wrangle Island in the Arctic Ocean, about 120 miles off the coast of northeast Siberia. They reached at most 70 percent of the size of their Siberian counterparts and lived until about 4,000 years ago. Dwarf forms of many other large animals also existed on other late ice age islands.

An unusual collection of now-extinct dwarf and giant animals occupied many Mediterranean islands from 1.6 million to 10,000 years ago. Before the end of the last ice age, humans occupied the island of Cyprus in the eastern Mediterranean. The landing of people on the island coincided with the disappearance of the pygmy hippopotamus, which was about the size of a small pig and roamed freely on Cyprus and other nearby islands. One site on Cyprus contained a large number of pygmy hippopotamus skull bones along with rock flakes supposedly made by human hunters, who lived on the island prior to 10,000 years ago. Apparently, this was a refuse pile, suggesting that people had a hand in the extinction of the pygmy hippopotamuses and possibly other animals as well.

Animals occupying islands often develop unique characteristics apart from their mainland relatives. Island birds are frequently flightless because they no longer need to take to the air to escape from predators. Historically, humans have accounted for the extinction of many island species. Often, species became extinct because of the destruction of their habitats or the introduction of predators or competing species.

THE LITTLE ICE AGE

From the late 15th century to the middle 19th century, the world was in the grip of a Little Ice Age. Average yearly temperatures fell as much as 1 degree Celsius, precipitation increased, and glaciers that had been steadily retreating since the end of the last ice age suddenly began to advance. The cooling period of the Little Ice Age coincided with an increase in volcanic

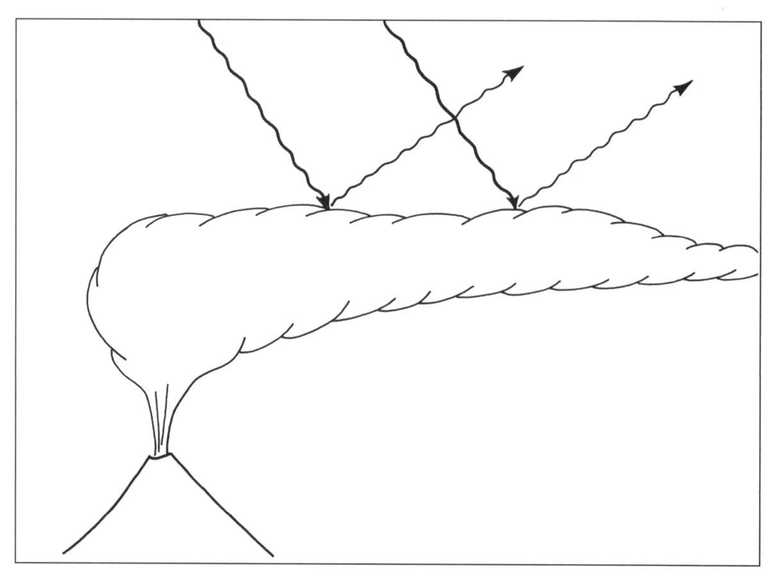

Figure 3–14 The effect of a volcanic dust cloud on incoming solar radiation.

activity. Volcanic ash in the higher atmosphere might have blocked out sunlight and lowered global temperatures (Fig. 3–14).

An apparent minimum of sunspot activity between 1645 and 1715 was linked to the coldest part of the Little Ice Age. The drop in sunspot activity might have been responsible for the unusually cold weather in Europe and North America. Ancient trees living during the same time and analyzed for their ring growth indicated an unusual climate during the Maunder Minimum of sunspot activity. The decrease in the sun's luminosity, or brightness, might have been as much as 1 percent, causing an estimated 2-degree cooling throughout the world.

During the relatively dry 15th and 16th centuries, major forest fires in North America appear to have occurred roughly once every 9 years. However, over the next three centuries, during the cool period of the Little Ice Age, forest fires were less frequent and less intense, occurring only about every 14 years.

The Norse, who had settled on the shores of Greenland around A.D. 980 and successfully raised cattle and crops, were decimated by the expanding ice sheets. The cooler climate brought bitter cold winters to northern Europe, and creeping glaciers chased people out of mountain valleys (Fig. 3–15), destroying farms and pasture. Forests decreased in size to be replaced by grassland and tundra, the snow line lowered appreciably, and glaciers grew in the Alps and other mountain ranges, extending farther down valleys. Cold winds blowing across the glaciers swept over the

Figure 3–15 The terminus of the Mer de Glace Glacier, Chamouni Valley, France-Switzerland, showing terminal and lateral moraines about the lower end of the glacier. Photo by J. A. Holmes, courtesy of USGS

foothills and plains. The long-lasting colder seasons ruined crops and harvests.

In the late 18th century, when American colonists fought Great Britain during the Revolutionary War, the severe winters threaten the colonial army almost as much as the British. George Washington's army nearly froze at Valley Forge. The extreme cold made it possible for the American forces to haul heavy cast-iron cannons across the thick ice of Long Island Sound, which seldom freezes over. Similarly, rivers and canals in Europe were frozen and impassible.

After the war, Benjamin Franklin was sent to Paris, France, to become the first diplomatic envoy from the newly created United States of America. During the summer, Franklin noticed a persistent dry fog over Europe and

North America. The haze blocked sunlight and substantially reduced surface temperatures. As a result, the winter of 1783–84 was among the most severe on record. For the first time in memory, normally free-flowing rivers in England and Europe froze over.

Franklin attributed the strange weather phenomenon to volcanic ash in the atmosphere from the explosive eruption of the Icelandic volcano Laki, which locally killed 10,000 people, roughly one-quarter of the total population, and 200,000 livestock. The volcanic ash spread across the Northern Hemisphere during the summer of 1783 and blocked the warming rays of the sun.

Icelandic eruptions apparently were more frequent during warm interglacials, when Iceland was largely free of ice. At the end of the last ice age, when Iceland lost its mile-high blanket of ice, the melting of that heavy load decreased the pressure on the Earth's mantle, making it easier for the solid mantle rock to melt. The glaciers apparently keep the volcanoes quiet by bearing down on the Earth's crust, which puts weight on the magma chambers that feed the volcanoes.

Volcanoes appear to be most active immediately after an ice age. When the ice melts, the pressure on the magma chambers is lifted, causing the volcanoes to erupt. Sediments from the bottom of the Norwegian Sea contain four layers of volcanic ash from Icelandic eruptions during the last 300,000 years. The layers indicae times when intense eruptions spewed out huge amounts of volcanic material that settled on the ocean floor. All layers fell within the short interglacial periods that punctuate the longer ice ages.

The 1815 explosive eruption of the Indonesian volcano Tambora on the island of Sumbawa (Fig. 3–16) exceeded any known volcanic eruption

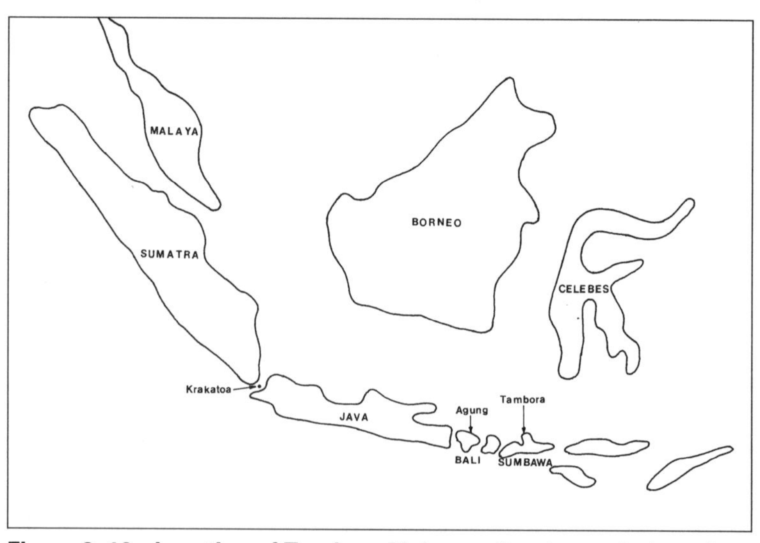

Figure 3–16 Location of Tambora Volcano, Sumbawa, Indonesia.

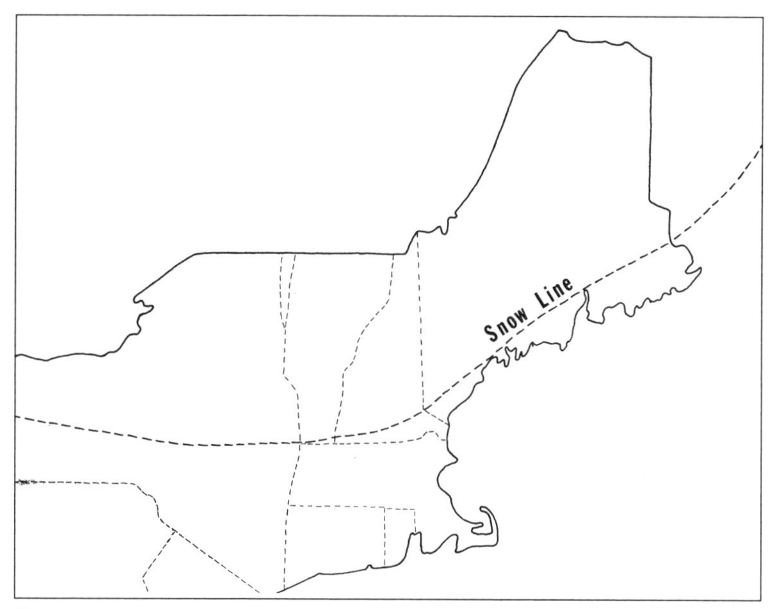

Figure 3–17 The snow line in New England in June 1816.

during the entire Holocene. It sent more volcanic ash into the upper atmosphere and obscured sunlight more than any volcano in the past 400 years. The eruption blew off the upper two-thirds of the mountain and cast some 25 cubic miles of debris into the atmosphere, which had a major impact on the climate throughout the Northern Hemisphere.

By the following summer of 1816, the ash had completely encircled the Earth, dropping temperatures as much as 3 degrees in New England and 2 degrees or more in Europe. Ship captains in the North Atlantic reported sighting large numbers of icebergs from a massive outbreak of Arctic ice. In New England, spring was late and snows were slow in melting (Fig. 3–17). When spring finally arrived and crops were beginning to grow, a killing frost in June took all but the hardiest plants. In autumn when harvest was about to begin, a cold wave out of the north brought widespread killing frosts and finished off crops that managed to survive the ordeals of summertime. The event went down in history as the "year without a summer."

The Europeans were much worse off because many regions were ravaged by the Napoleonic wars that ended the year before. In parts of England, temperatures dropped 3 degrees below normal, and rain fell on all but a few days from May to October. Crops were killed by frost, failed to ripen, or rotted in the field. The potato crop failed in Ireland, causing massive famine, and food riots occurred in Wales. The scarcity of food brought on insurrections and riots and eventually disease in other parts of Europe that killed over 100,000 people.

4

CAUSES OF GLACIATION

Any theory that attempts to explain the waxing and waning of the ice ages about every 100,000 years must demonstrate how massive glaciers developed and galloped across the continents and why they abruptly retreated back to the poles. Once the ice sheets were in place, they appeared to control the climate to maintain their own existence by reflecting solar energy out into space. The dryer climate spawned huge dust storms that shaded the Earth and kept it cool, ideal conditions for the continuation of an ice age.

With such powerful feedback forces in place, the great ice sheets would seem to be able to maintain themselves indefinitely. Yet the geologic record shows that dramatic changes in the climate did occur, causing the glaciers to collapse. The challenge is finding a mechanism that regularly altered the climate over the last several million years, turning the ice ages on and off almost like clockwork.

GALACTIC DUST CLOUDS

The Solar System revolves once around the Milky Way Galaxy about every 200 million years. Two or three times a century, a giant star explodes and becomes a supernova. The expanding supernova injects massive amounts of gas and dust into the galaxy (Fig. 4–1). The Solar System enters such a dust cloud possibly every 100 million years or so. When passing through relatively dense regions of the intergalactic dust cloud, the material falling into the sun could lower its output. The dust also might affect the Earth's climate by forming ice clouds in the upper atmosphere that block out sunlight.

The passage through the dust cloud could take several million years and might have been responsible for earlier glacial periods, which lasted for a similar length of time. However, such a journey is too lengthy to explain the relatively short ice ages of the past 3 million years. Furthermore, the transit through a dusty arm of the galaxy could not be responsible for the continuous temperature decrease since the Cretaceous, the warmest period of the last 600 million years.

A cloud of interstellar gas does appear to be streaming through the Solar System, and in its lifetime the sun has passed through at least 100 dense gas clouds. During the 100,000-year traverse through the interstellar gas cloud, the Earth acquires a large amount of molecular hydrogen. The hydrogen reacts with chemical constituents in the high reaches of the

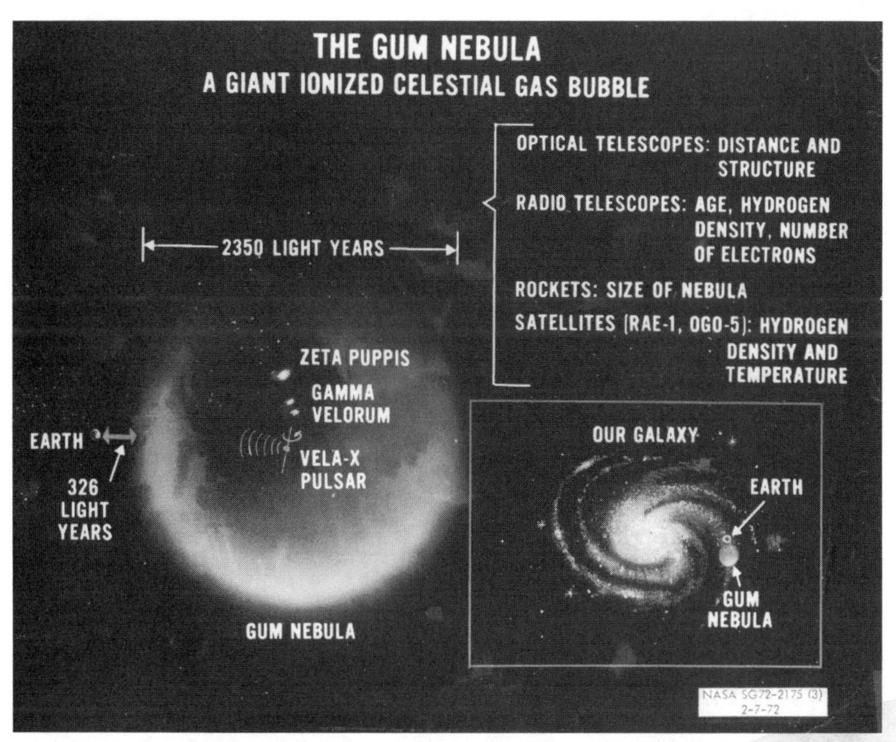

Figure 4–1 The Gum Nebula was probably produced when the burst of radiation from a supernova heated and ionized gas in interstellar space. Courtesy of NASA

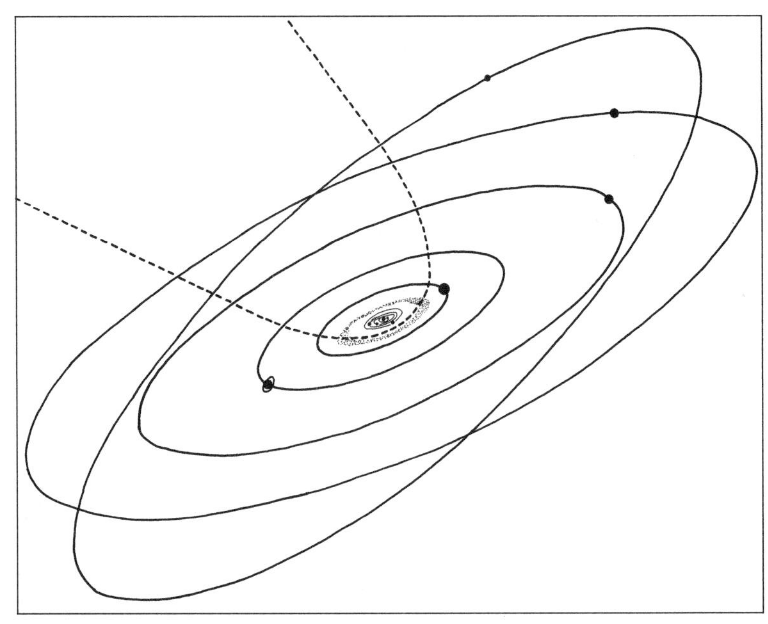

Figure 4–2 Comets travel in oblique orbits with respect to the Solar System.

atmosphere to produce water vapor, which condenses into clouds. The clouds reflect solar radiation, substantially lowering surface temperatures.

The sun's journey through dense clouds located at the midplane of the galaxy also could reduce the Earth's insolation and initiate climatic changes. But the dust cloud does not appear to be dense enough to significantly block out the sun during each passage through the midplane, which could take upward of several million years. However, when the sun reaches the upper or lower regions of the galaxy, the Earth might be exposed to higher levels of cosmic radiation from supernovas, which could ionize the upper atmosphere and produce a haze that blocks out sunlight and cools the planet.

As the Solar System passes through the galactic midplane roughly every 30 million years, the increased gravitational attraction might distort the Oort Cloud, a shell of over a trillion comets about a light-year away. The distortion in the gravity field might hurl thousands of comets crashing into the inner Solar System (Fig. 4–2), many of which could strike the Earth, and the ensuing bombardment would cause environmental havoc.

SOLAR RADIATION

For centuries, astronomers have referred to a solar constant, whereby the total amount of solar energy impinging on the Earth has remained steady through time. The solar constant depends on the sun's luminosity, or brightness, and the Earth's orbit. The luminosity depends on the sun's size and surface temperature. A reduction of the solar constant by only a few percent is sufficient to initiate an ice age.

Long-term changes in luminosity are not detectable from the Earth or by orbiting satellites. However, small, short-term fluctuations sufficiently large to produce variations in the climate are detected. These minor changes in the solar output occur in regular intervals of 22 years, known as the solar cycle. Coinciding with the solar cycle is an 11-year sunspot cycle. Cycles of 90 and 180 years also occur, and longer-period solar cycles might correlate with the ice ages.

During a sunspot maximum, when large numbers of sunspots mar the sun's surface (Fig. 4–3), its activity increases. During times when few sunspots exist, the sun cools by as much as 1 percent. Such a condition might have existed during the Little Ice Age, a span of unusually cold weather in Europe and North America from about 1500 to 1850.

The Maunder Minimum was a 70-year cessation of sunspot activity from 1645 to 1715 (Fig. 4–4). It correlates with the coldest part of the Little Ice Age when global temperatures dropped 1 degree Celsius and glaciers were

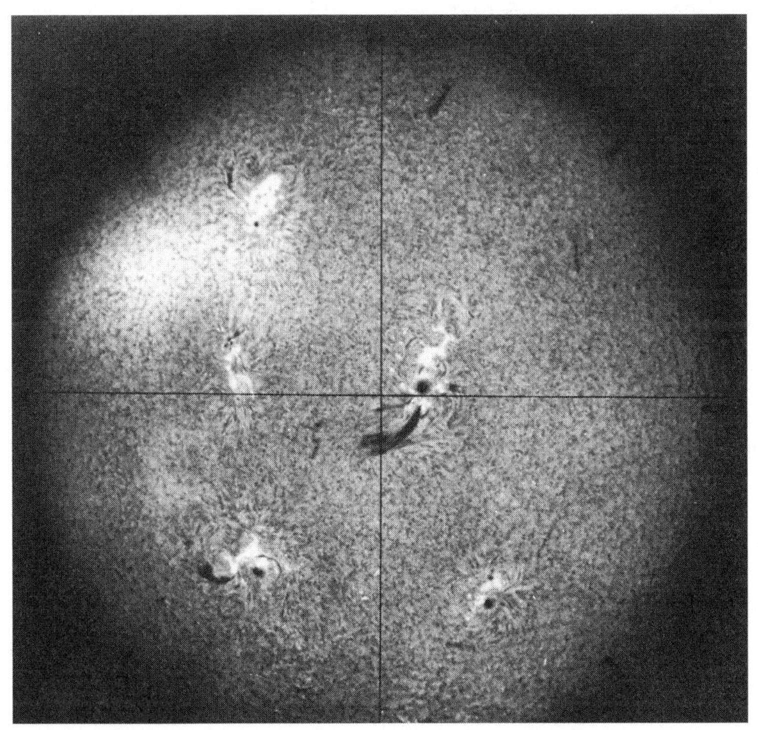

Figure 4–3 Sunspots and other solar activity viewed from the solar telescope on board *Skylab*. Courtesy of NASA

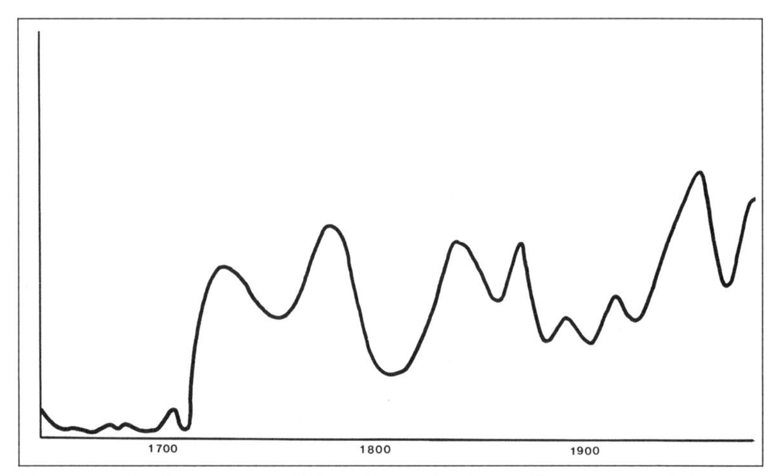

Figure 4–4 Sunspot activity through time. Note the minimums between 1645 and 1715.

again on the advance. The unusual low level of solar activity is also supported by a gap in Chinese naked-eye sunspot records for the same period. Tree rings formed during this time indicated a period of very unusual behavior of the sun, matching closely that of the Maunder Minimum. Another Maunder-type minimum occurred during the Ming dynasty in China between 1400 and 1600, which might have sparked the Little Ice Age.

The solar cycle also might be regulated by the gravitational forces of the inner planets and Jupiter. Throughout this century, the alignment of the inner planets on one side of the sun appeared to produce different sunspot numbers. Jupiter's orbital period happens to be roughly the same as the 11-year sunspot cycle, and since Jupiter is the largest planet it would have the greatest gravitational affect on the sun.

When the planets line up on the same side of the sun, their gravitational pull raises tides on its surface like the moon does on the ocean. The amplitudes of the tides are extremely small due to the greater distance and less gravitational attraction of the planets. However, the years of sunspot maximum and minimum since 1800 coincide closely with the sun's tidal maximum and minimum.

The alignment of the planets also might directly influence the Earth's weather. The ancient Chinese astronomers were the first to recognize this alignment called a planetary synod. Chinese researchers found that planetary alignments have affected the weather for the past 3,000 years. The 180-year synods also coincide with the variations in the temperature record of the Greenland ice cores and match the 180-year cycle of solar activity.

About every 180 years, the Earth is on one side of the sun while the other planets reside on the other side. This displaces the gravitational center of the Solar System and stretches the Earth's orbit nearly 1 million miles.

Because it is farther from the sun, the Earth cools by a small amount for several years. The last planetary synod began in October 1982, which could initiate a cold spell lasting nearly half a century provided other influences such as man-made greenhouse warming do not become a factor.

ORBITAL MOTIONS

All possible orbital motions are identified by the geometry of the Earth's orbit, the precession of the equinoxes, and the tilt of the rotational axis. Changes in these orbital elements, called Milankovitch cycles (Fig. 4–5) for the astronomer Milutin Milankovitch, do not affect the total amount of yearly insolation but only alter the amount of solar radiation reaching specific latitudes in certain seasons. One region might experience cold winters and hot summers during one cycle and mild winters and cool summers during another cycle. Therefore, cooler summers when snow and ice fail to melt can initiate an ice age.

The Earth's orbit around the sun alternates from nearly circular to elliptical about every 100,000 years. This orbital motion is called the eccentricity cycle. While in a circular orbit, the Earth maintains a constant distance of 93 million miles from the sun during all seasons, and the amount of insolation remains the same throughout the year. During the ellipse phase, the Earth is closer to the sun in one season, making it warmer, and farther from the sun in the opposite season, making it cooler.

Presently, the Earth has an elliptical orbit, and its perihelion (closest point to the sun) occurs in early January and its aphelion (farthest point

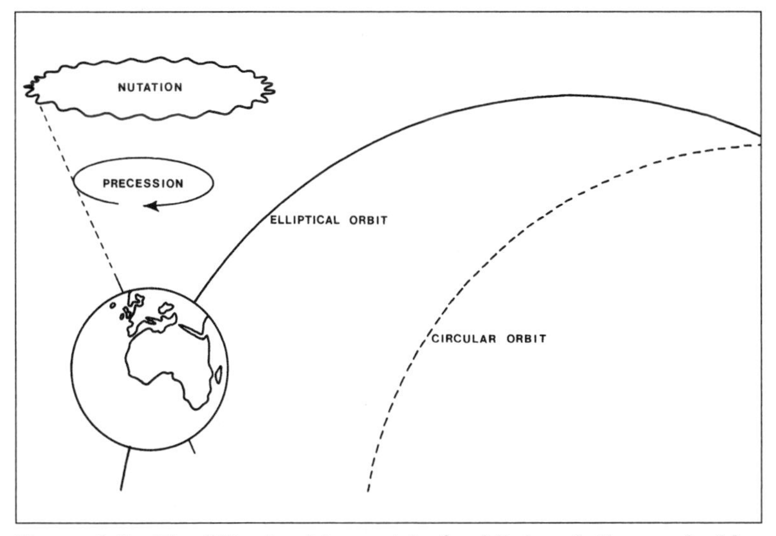

Figure 4–5 The Milankovich model of orbital variations coincides with the ice age cycles.

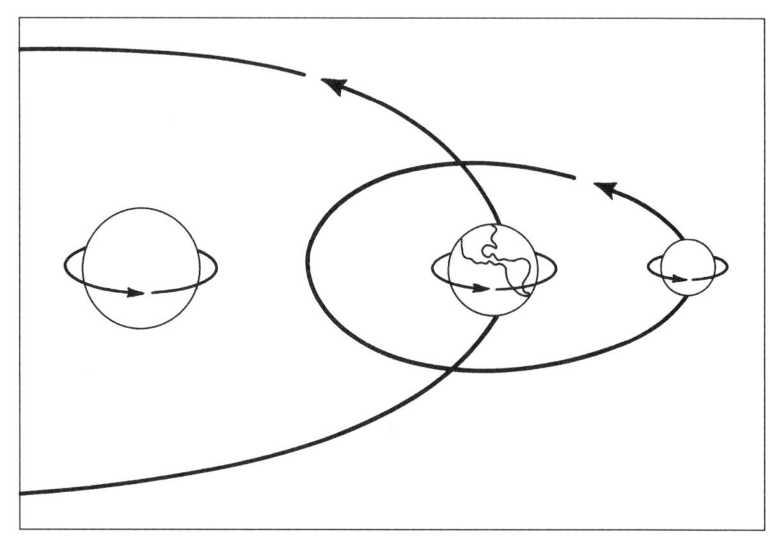

Figure 4–6 Orbital motions of the sun, Earth, and moon. The sun completes one rotation on average in 27 days. The moon's rotation and revolution around the Earth is 27.5 days, and the Earth's rotation rate is 24 hours.

from the sun) occurs in early July, which makes winters in the Southern Hemisphere slightly cooler than northern winters. The difference between perihelion and aphelion is about 3 million miles, and the total insolation is 7 percent less during the northern summer than in winter.

This orbital cycle might explain the recurrence of the Pleistocene ice ages every 100,000 years over the last million years or so. Yet it remains a mystery why the weakest cycle, responsible for less than 1 percent of the total variation of insolation, produces the largest changes in climate. Perhaps it only initiates climatic changes that amplify and reinforce each other.

In contrast, the shorter tilt and precession cycles alter the amount of sunlight the Northern Hemisphere receives in summer by as much as 20 percent. The tilt cycle has its strongest impact on high-latitude sunlight, while the precession cycle affects sunlight in the tropics. The Earth's rotational axis tilts at an angle of 23.5 degrees to the plane of its orbit, called the ecliptic. The sun and moon exert a gravitational pull on the spinning Earth (Fig. 4–6), making its axis precess, or wobble, like a toy top.

The axis thus describes a cone in the heavens as it precesses clockwise, or in the opposite direction of the Earth's rotation. One precessional cycle takes about 23,000 years, which means that around 10,000 years ago, at the end of the last ice age, Vega was the North Star instead of Polaris, and the seasons were reversed. In 10,000 years hence, the Earth will again be tilted in the opposite direction, with winter replacing summer.

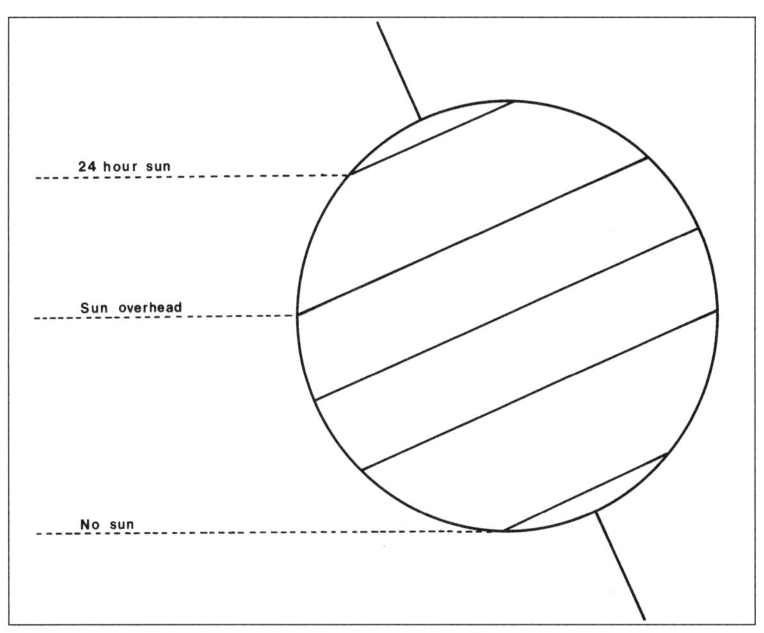

Figure 4–7 The effect of the tilt of the Earth's axis on the seasons. Shown is summer in the Northern Hemisphere.

Over the course of geologic history, the tilt of the Earth's axis has varied from 21.5 to 24.5 degrees. The effect of altering the angle of tilt is to shift the position where the sun is overhead during the seasons (Fig. 4–7). The greater the tilt, the larger the difference between summer and winter temperatures. If the Earth's axis was not tilted and perpendicular to the ecliptic, the planet would have no seasons.

When the axis is steeply tilted, the Earth experiences extremely large temperature variations from one season to the next. Even slight changes in the degree of tilt can cause major climatic effects. The Earth completes one full tilt cycle about every 41,000 years, and since the end of the last ice age the degree of tilt has been steadily decreasing, with the prospect of cooler summers and warmer winters.

The variations in orbital motions work in concert to produce the overall changes in the pattern of solar radiation impinging on the Earth. They can combine in such a manner as to bring about the worst possible climatic conditions that might lead to glaciation. If the amount of summer sunshine in the Northern Hemisphere dropped, causing cool summers, the volume of the ice sheets would grow in proportion to the sunlight deficit. The previous winter's snow might fail to melt, and the following winter's snow would accumulate into thick ice fields. If this condition continued unabated for several years, the whole process of changing from an interglacial to a full ice age could occur in as little as a century.

The Milankovitch model cannot be proven without establishing reliable dates for the ice ages. One means is by dating the fluctuations in sea level. During a glacial period, the ice sheets lock up large quantities of the Earth's water and the ocean lowers appreciably. Since corals only live in shallow water (Fig. 4–8), fluctuations in sea level leave a terrace of coral growth corresponding to periods of glaciation. Radiometric dating techniques determine the age of the ancient coral reefs. The coral terraces date some 20,000 years apart, comparing favorably with the precessional cycle of the Earth's axis.

The Deep Sea Drilling Project sponsored by the National Science Foundation has yielded evidence for 400,000-, 100,000-, and 40,000-year cycles by analyzing the calcium carbonate content of ocean bottom sediments. Climate-related changes in the dissolving power of seawater, in the level of the sea, in circulation patterns, and in the erosion rate of the continents affect the proportion of calcium carbonate on the ocean floor. Tiny marine organisms use the carbonate to make their shells. When they die and are buried in the sediments, the composition of their shells indicate the climate during their lifetime.

Figure 4–8 Cocos Island, Guam, showing coral reef and lagoon. Photo by K. O. Emery, courtesy of USGS

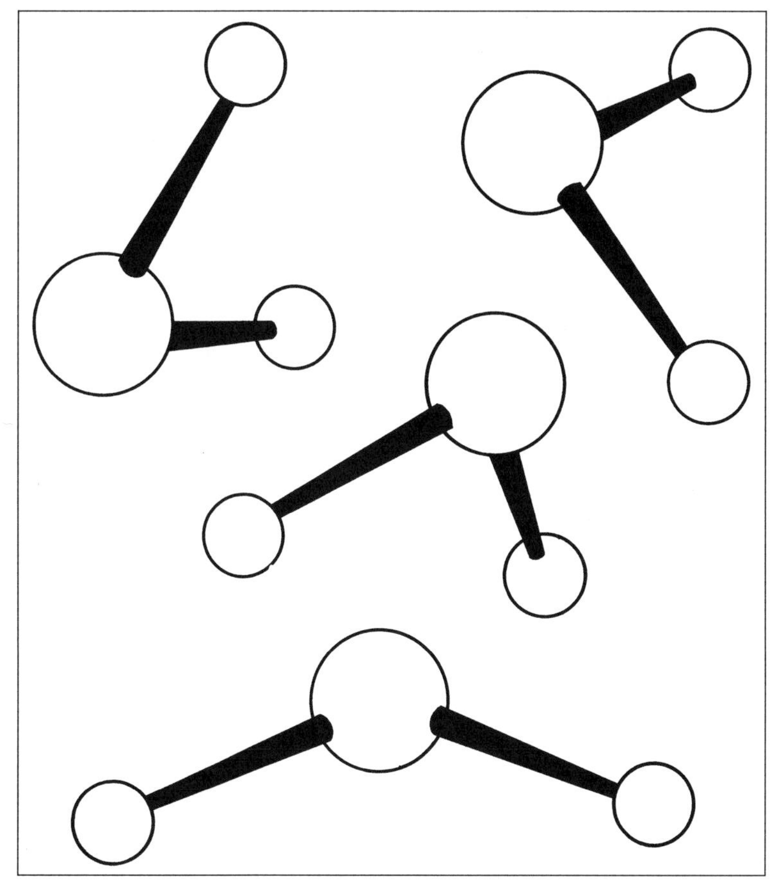

Figure 4–9 As the temperature lowers, lighter water molecules evaporate more easily than heavier ones.

A comparison of the content of the light oxygen isotope O_{16} and the heavy oxygen isotope O_{18} in the fossilized shells of these creatures provides a direct means for determining the Earth's climate. As seawater evaporates during a warm climate, water molecules composed of both oxygen isotopes (Fig. 4–9) rise into the atmosphere and precipitate as snow at the poles. However, during cooler climates, water molecules weighed down by O_{18} isotopes are left behind and become concentrated in the seawater. Therefore, the more O_{18} locked up in the fossil sells of marine organisms the colder the climate.

Chemical analysis of the Greenland and Antarctic ice cores has yielded another method of temperature measurement by comparing the ratios of O_{16} to O_{18}. One striking feature of this study is how clearly it follows the 41,000-year tilt cycle, yielding strong support for the Milankovitch model of glaciation.

In a flooded fissure in Nevada called Devils Hole, about 40 miles east of Death Valley, are layers of calcite that contain information on rainfall changes in the area since about 300,000 years ago. The data were obtained by analyzing the oxygen isotope ratios within the calcium carbonate similar to that used on marine sediments. The changes recorded at Devils Hole hinted of cycles lasting 100,000, 40,000, and 23,000 years that matched those of the orbital cycles, suggesting that these variations controlled the timing of the climatic changes in the area.

The Earth's orbital variations might have modulated the climate for hundreds of millions of years. In banded lake bed sediments 200 million years old from the Newark Basin in northern New Jersey are repetitive sequences that correspond to cycles of varying lake depths. The fluctuations in water level are influenced by the changing distribution of sunlight over the globe during any given season. The cycles closely resemble the periods of precession and tilt of the Earth's axis and the eccentricity of its orbit.

Dating the sediments yielded periods of 25,000, 44,000, 100,000, 125,000, and 400,000 years. These compared reasonably well with the present orbital periods of 23,000, 41,000, 95,000, 123,000, and 413,000 years. Similar cycles were verified in sediments throughout the world. Therefore, the orbital variations appear to have operated on the climate throughout geologic time.

GEOGRAPHICAL INFLUENCES

The positions of the continents determine global climatic conditions (Fig. 4–10). The movement of continents around the world by mobile crustal plates greatly influenced global temperatures, ocean currents, biological productivity, and many other factors of fundamental importance to the Earth. When most of the land huddled near the equatorial regions, the climate was warm. However, when lands wandered into the polar regions, they became ice covered. Furthermore, during times of highly active continental movements, greater volcanic activity occurs, which could affect the composition of the atmosphere, the rate of mountain building, and ultimately the climate.

The gathering of continents around the polar regions during the last few million years has resulted in extended periods of glaciation because high-latitude land has a higher albedo and lower heat capacity than the surrounding seas, which encourages the accumulation of snow and ice. With more land residing in the higher latitudes, the colder and more persistent is the ice, especially when much of the land is at higher elevations, where glaciers grow easily.

Replacing land in the tropics with ocean also has a net cooling effect because land in the tropics absorbs more solar radiation, while the oceans

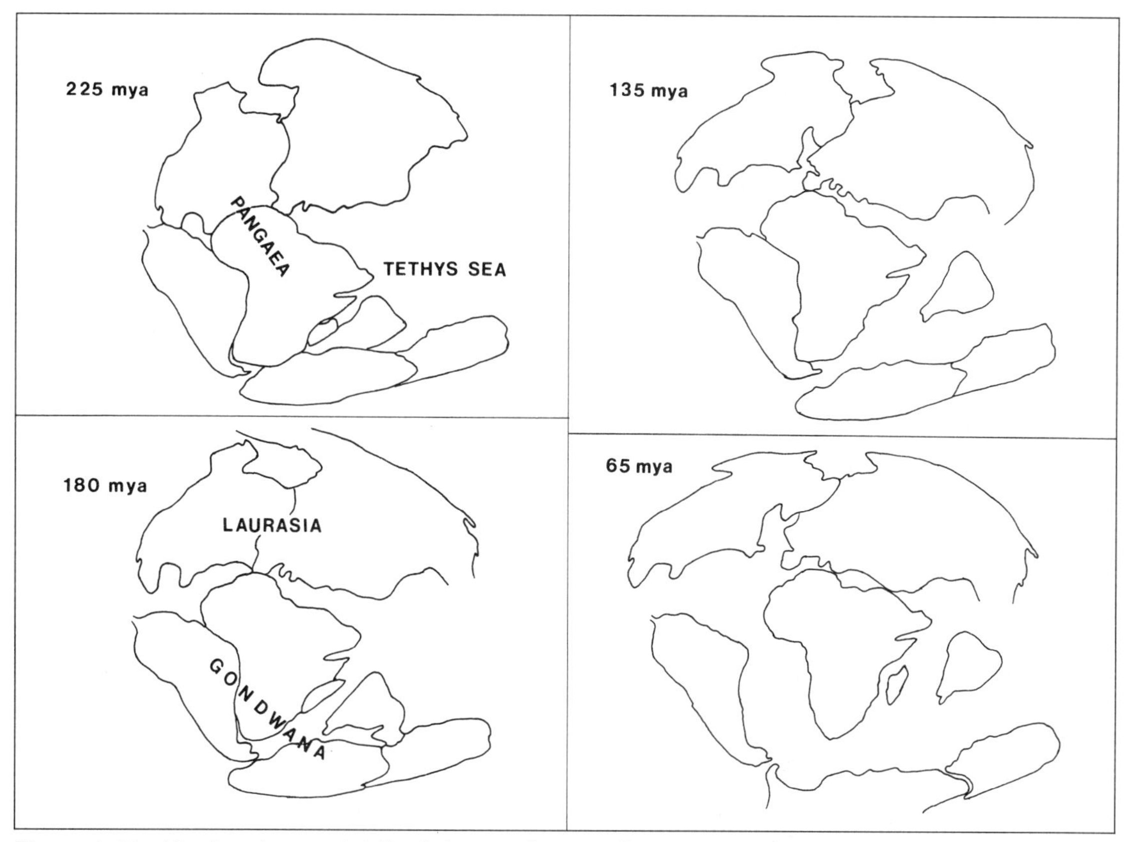

Figure 4–10 The breakup and drift of the continents affected climatic conditions.

reflect sunlight back into space. Increasing the land area in the high latitudes and allowing snow to cover it results in a permanent polar glacial climate.

Land congregating in one area also affects the shapes of the ocean basins. The ocean bottom influences how much heat ocean currents carry from the tropics to the poles. When Antarctica separated from South America and Australia and moved over the South Pole some 40 million years ago, the polar vortex established a unique circumpolar Antarctic current that isolated the frozen continent and prevented it from warming by poleward flowing waters originating in the tropics.

MOUNTAIN BUILDING

Changing climatic patterns during the Cenozoic era resulted from the movement of continents toward their present positions along with intense tectonic activity that built landforms and raised most mountain ranges of the world. The period is remarkable for its intense mountain building, and a spurt in mountain growth over the past 5 million years, during which the rate of uplift has more than doubled, might have helped trigger the Pleistocene ice ages.

During mountain-building episodes, great blocks of granite soared high above the surrounding terrain (Fig. 4–11). In the higher mountain reaches, glaciers developed when the Earth suddenly turned cold during the Pleistocene epoch. Land rising to higher elevations where temperatures are colder allows glaciers to grow, especially in the higher latitudes.

Prior to 40 million years ago, most of the world was warmer and wetter than today. India, which broke away from Antarctica about 130 million years ago, slammed into southern Asia, and the increased buoyancy uplifted the Himalaya Mountains and the broad Tibetan Plateau, whose equal has not existed on this planet for over a billion years. The continental collision heated vast amounts of carbonate rock, spewing several hundred trillion tons of carbon dioxide into the atmosphere, which might explain why the Earth grew so warm during the early Cenozoic.

In the past 40 million years, and especially the last 15 million years, the warm, wet climate largely disappeared and was replaced with colder climates, producing much greater regional extremes in precipitation. During the past 5 to 10 million years, the Himalayas and adjoining Tibetan Plateau rose over a mile in elevation, and other mountain ranges were similarly affected.

Over the last 3 million years, the Earth grew so cold it began to experience periodic ice ages. The culprit appears to be a spasm of geologic upheaval,

Figure 4–11 An alpine glacier in the Alaskan Range, Copper River region, Alaska.
Photo by W. C. Mendenhall, courtesy of USGS

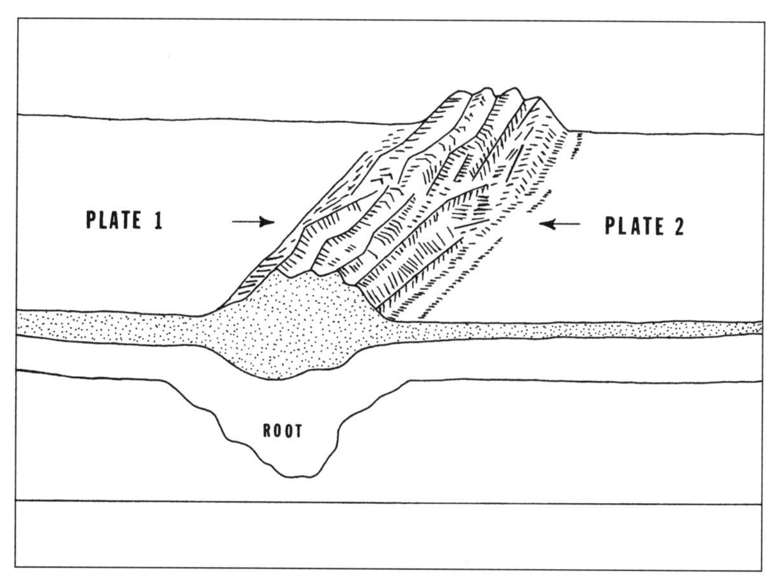

Figure 4–12 Continental collisions forced up mountain ranges at the point of contact between two crustal plates.

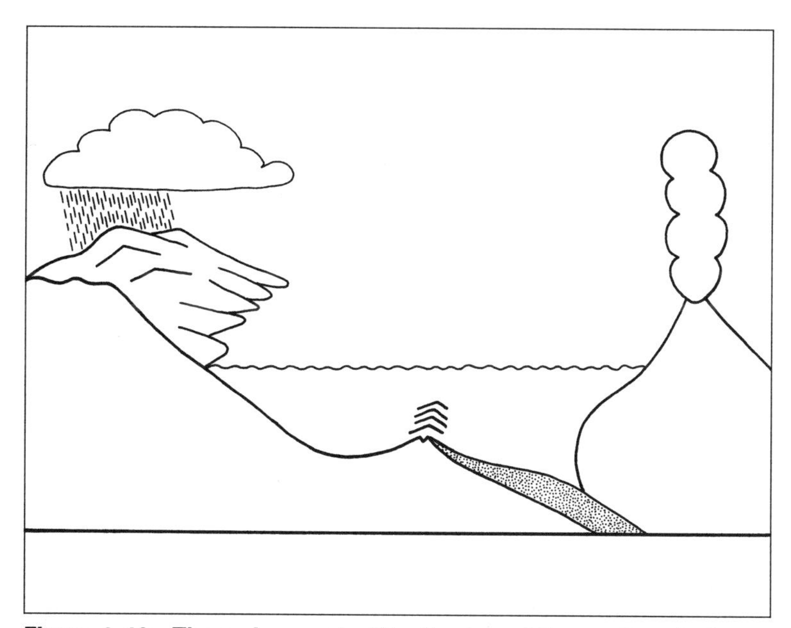

Figure 4–13 The carbon cycle: Weathering of the continents draws carbon dioxide from the atmosphere and converts it into bicarbonate, which washes off the land into the ocean, where carbonate sediments melt in the Earth's interior and volcanoes return carbon dioxide to the atmosphere.

resulting in the development of huge elevated regions in several parts of the world, particularly in southern Asia with the elevation of the Himalayas and the Tibetan Plateau, western North America with the raising of the Rocky Mountains, and western South America with the rising Andes Mountains. The dramatic increases in elevation of these regions significantly affected the physical and chemical properties of the atmosphere, which helped shape the present climatic trends. The high plateaus diverted the prevailing wind flow and created thermal effects that produce powerful cyclonic pressure systems.

Mountain building associated with the movement of crustal plates (Fig. 4–12) elevated massive chunks of crust, where glaciers are nurtured in the cold, thin air. Land rising to higher elevations spurs the growth of glacial ice, especially in the higher latitudes. Moreover, continents scattered around the world interfere with ocean currents, which distributed heat over

Figure 4–14 Alpine glaciation by latitude. White areas represent glaciated regions and stippled areas represent the extent of forest growth.

the globe. Mountain building also alters patterns of river drainages and climate, which in turn affects terrestrial habitats. The higher topographies could have disturbed the jet streams in the Northern Hemisphere, diverting frigid Arctic air to the south.

The elevated landforms also increased the weathering of rock materials because uplift continuously exposes fresh rock to the weather, and water is more erosive spilling down a mountainside than flowing across flatland. The increased weathering drew carbon dioxide out of the atmosphere and deposited it as carbonates at the bottom of the ocean (Fig. 4–13), significantly weakening the greenhouse effect that warms the Earth. Normally, volcanic activity on the ocean floor and on the continents balances out the carbon dioxide deficit. But over the past 5 million years, these restorative processes have not kept pace with the depletion of carbon dioxide caused by increased weathering.

Glaciers might have formed and persisted on continents even at low latitudes as long as they maintained a high elevation. This is because temperature decreases with altitude, and every 1,000 feet of elevation results in an equivalent increase of 300 miles of latitude, so that the top of a 20,000-foot mountain located at the equator would be as frigid as the polar regions (Fig. 4–14). By the time the continents had wandered to their present positions and all the mountain ranges had risen to their current heights, the world was ripe for the ice ages.

THE CARBON CYCLE

The level of carbon dioxide in the atmosphere played a major role in the progression and regression of the ice ages. Soviet scientists have recovered an ice core that samples most of the 12,000-foot ice cap beneath their Vostok Station on the Magnetic South Pole in East Antarctica. The core contains a continuous record of the temperature and composition of the atmosphere for the past 220,000 years, spanning two glacial and two interglacial periods.

Air bubbles trapped in the ice provided information on the carbon dioxide content of the atmosphere at the time the ice was laid down, while deuterium (an isotope of hydrogen) revealed the temperature. What was highly significant is that the level of carbon dioxide and the temperature kept pace throughout the period (Fig. 4–15).

During the last ice age, the level of atmospheric carbon dioxide was about 0.02 percent, or roughly half today's value. Moreover, atmospheric methane, which is the second most important greenhouse gas contributing to about one-quarter of the warming, also decreased significantly. About 22,000 years ago, at the height of glaciation, the amount of atmospheric methane was roughly half the preindustrial level. The reduction of methane

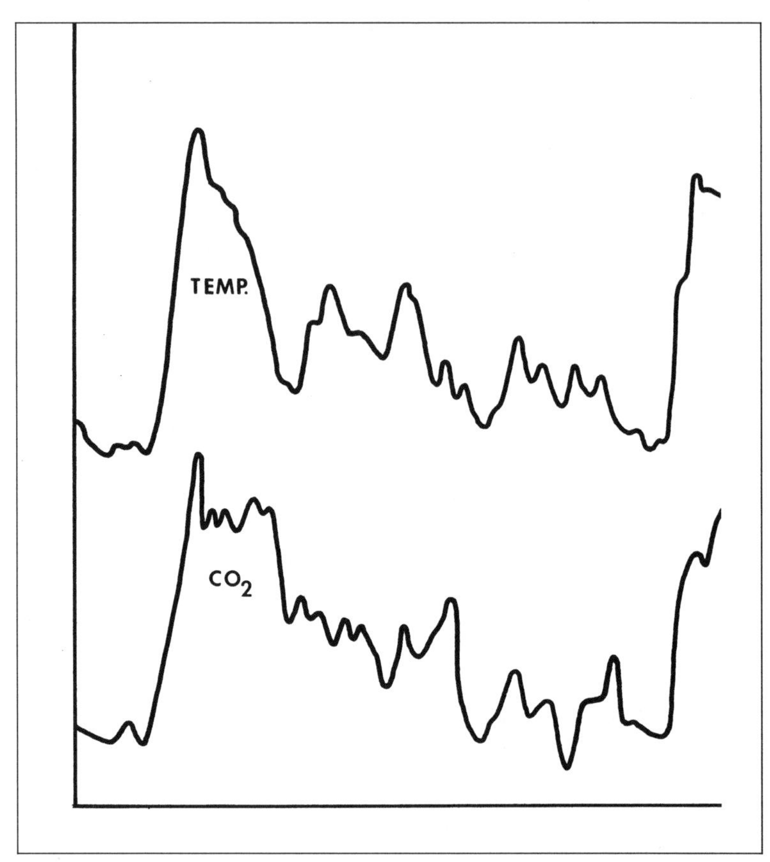

Figure 4–15 The global temperature (top curve) and atmospheric carbon dioxide (bottom curve) have kept pace for the last 160,000 years.

during the ice age is blamed on the lower biological activity of wetlands and other habitats due to a colder climate.

The creation and decomposition of peat bogs could have been responsible for most of the changes of atmospheric carbon dioxide during the past two glaciations. The bogs have accumulated upward of 250 billion tons of carbon in the last 10,000 years since the end of the last ice age, mostly in the temperate zone of the Northern Hemisphere. Progressively more land has moved to latitudes where it stores large amounts of carbon as peat. Over the last million years, glaciations have gradually remodeled large parts of the Northern Hemisphere into landforms more suitable for peat bog formation in wetlands.

The ocean plays a critical role in drawing down the level of atmospheric carbon dioxide. In the upper layers of the ocean, the concentration of gases is in equilibrium with the atmosphere. The gases dissolve into the waters of the ocean mainly by agitation of surface waves. If the ocean were lifeless, much of its reservoir of dissolved carbon dioxide would enter the atmosphere, more than tripling its present content.

Fortunately, the ocean is teeming with life, and marine organisms use carbon dioxide as dissolved carbonates to build their skeletons. When the organisms die, their skeletons sink to the bottom of the ocean, where they dissolve in the deep waters of the abyss, which holds by far the largest reservoir of carbon dioxide, 65 times more than the atmosphere.

On the shallow seafloor, the carbonate skeletons form deposits of carbonaceous sediments such as limestone (Fig. 4–16). The burial of carbonate in the crust is responsible for about 80 percent of the carbon deposited on the ocean floor. The rest of the carbonate derives from the burial of dead organic matter washed off the continents.

Half the carbonate transforms back into carbon dioxide, which escapes into the atmosphere. Without this process, in a mere 10,000 years, all carbon dioxide would have been removed from the atmosphere, ending photosynthesis and life. Even a reduction of half the present amount of carbon dioxide is enough to start a new ice age.

In this respect, marine life acts like a pump that removes carbon dioxide from the ocean's surface and the atmosphere and stores it in the deep sea.

Figure 4–16 A limestone formation of the Bend Group in a ravine at the base of the Sierra Diablo escarpment, Culberson County, Texas. Photo by P. B. King, courtesy of USGS

TABLE 4–1 AMOUNT OF CARBON RELATIVE TO LIFE

Source	Relative Amount
Calcium carbonate in sedimentary rocks	60,000
Ca-Mg carbonate in sedimentary rocks	45,000
Sedimentary organic matter in the remains of animal tissues	25,000
Bicarbonate and carbonate dissolved in ocean	75
Coal and petroleum	7
Soil humus	5
Atmospheric carbon dioxide	1.5
All living plants and animals	1

The faster this biological pump works, the more carbon dioxide that is removed from the atmosphere. This rate is determined by the amount of nutrients in the ocean, which respond to changes in ice volume.

When the ice sheets began melting about 14,000 years ago, atmospheric carbon dioxide rapidly increased until about 10,000 years ago. The rise in

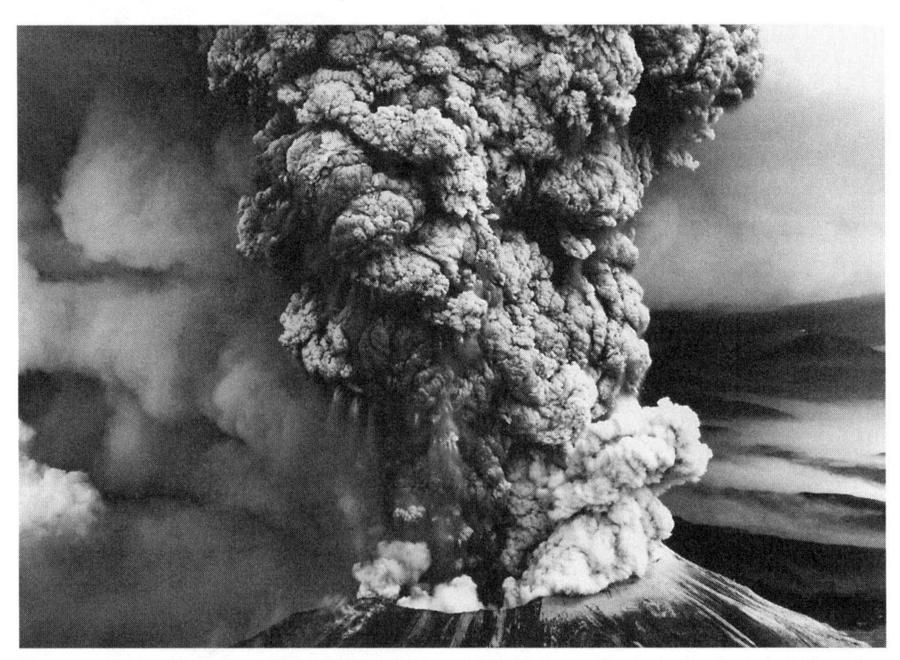

Figure 4–17 Eruption cloud from Mount St. Helens during the height of major eruptive activity on May 18, 1980. Photo by A. Post, courtesy of National Forest Service

sea level from the influx of meltwater flooded continental shelves, which led to the removal of organic carbon and nutrients. With a reduction of nutrients, the biological pump slowed, allowing deep-sea carbon dioxide to return to the atmosphere.

Increased volcanic activity toward the end of the ice age played another important role in restoring the carbon dioxide content of the atmosphere. One of the most important volatiles in magma is carbon dioxide, which helps make it flow easily. The carbon dioxide escapes from sediments when they melt in the Earth's interior. The molten magma along with its content of carbon dioxide rises to the surface to feed volcanoes that lie on the edges of subduction zones and at midocean ridges. An erupting volcano releases tremendous amounts of volcanic ash and carbon dioxide into the atmosphere (Fig. 4–17), which completes the carbon cycle.

VOLCANIC ERUPTIONS

A massive volcanic outburst on the Kamchatka Peninsula in the North Pacific closely coincides with an abrupt descent into the Pleistocene Ice Age about 2.6 million years ago, when northern hemispheric glaciation greatly intensified. Volcanic ash layers on the bottom of the ocean were 5 to 10 times more abundant in the Pleistocene than during most of the past 20 million years. The burst of volcanism might have been little more than an extra burden to an already deteriorating climate system. For tens of millions of years, the Earth had been sliding toward more cold and ice. Then for unknown reasons, it suddenly leaped into the ice age.

The amount of volcanism could affect the composition of the atmosphere, the rate of mountain building, and the climate. A large number of volcanoes erupting over a long interval could lower global temperatures by injecting huge amounts of volcanic ash and dust into the upper atmosphere. Heavy clouds of volcanic dust have a high albedo and reflect solar radiation back into space (Fig. 4–18), thereby shading the Earth and lowering global temperatures. A 5-percent reduction in solar radiation reaching the Earth's surface could result in a drop of global temperatures as much as 5 degrees, enough to initiate an ice age.

The eruption cloud contains particles, ranging in size from coarse ash to dust. The particles that remain airborne for long periods have the largest effect on the climate, depending on the nature of the dust and its location in the atmosphere. Volcanoes eject ash and dust into the troposphere and lower stratosphere up to altitudes of 20 miles as well as finer particles to altitudes of 30 miles and more. The volcanic particles in the lower altitudes seem to have the greatest influence on the climate, producing dense, long-lived dust clouds.

Volcanoes also erupt vast quantities of water vapor and gases, including sulfur dioxide, which reacts with water to produce sulfuric acid. These

Figure 4–18 The volcanic ash cloud from the 1980 eruption of Mount St. Helens.
Courtesy of NOAA

aerosols might penetrate the stratosphere like a fine mist and obscure sunlight. Aerosols move swiftly eastward around the globe but their spread into the higher latitudes is relatively slow. They are also transparent to outgoing infrared radiation, and this heat loss would further cool the Earth. With the combination of both dust and aerosols expelled into the atmosphere, large volcanic eruptions appear to make a significant impact on the climate.

The climatic effects of volcanic dust depend on the type and size of the dust particles and where they concentrate in the atmosphere. Large dust particles in the troposphere trap heat rising from the ground that would otherwise escape into space and thereby warm the Earth. Conversely, the smaller particles and aerosols tend to allow heat from the surface to escape into space, while blocking the sun's rays from reaching the ground.

Many volcanic eruptions occurring around the world in a short time span inject so much dust into the upper atmosphere that volcanoes alone could cause the onset of an ice age. This argument is supported by geologic evidence of thin layers of volcanic dust buried in sediments that correlate with times of increased ice cover. Ash layers in sediment cores from the Mediterranean Sea and the Indian Ocean indicated regular bursts of volcanic activity every 23,000 years, corresponding to the precessional cycle of the Earth's spin axis. Evidence found in Lake Biwa, Japan, suggest volcanic activity every 100,000 years correlating with the eccentricity cycle. Volcanic debris also exists in various layers of the Greenland and Antarctic ice cores, providing strong evidence that frequent and violent eruptions accompanied the last ice age.

The massive Toba eruption on Sumatra, about 73,500 years ago, was the largest explosive volcanic cataclysm of the last 2 million years. When the volcano blew its top, the space occupied by the mountain peak became the world's largest volcanic caldera filled with a 500-square-mile lake. Toba hurled some 10 billion tons of volcanic ash and gases 20 miles into the atmosphere. The debris cloud, which made the atmosphere more reflective to sunlight, could have reduced the amount of solar radiation striking the Earth's surface, dropping temperatures 3 to 5 degrees. The eruption might have accelerated global cooling when glaciation was already underway, providing an extra kick to the climatic system.

5

EFFECTS OF GLACIATION

Glaciation has had a major influence on the world's environment. The weight of the glaciers might have produced additional volcanic activity, and the subsequent ash and dust would have blocked out sunlight, cooling the planet. The lowered global temperatures might have resulted in many changes in the Earth's biology. Species forced to live in relatively warm, narrowly confined regions around the equator, would have to compete for space, causing mass extinctions.

The heavy glaciers might have caused the geomagnetic poles to wander because the extra weight generated instability on the spinning Earth. Large amounts of ice in the polar regions affect atmospheric and oceanic circulation, causing ocean currents to run "backward." But the most dramatic effect comes from locking up large quantities of water in the ice sheets, dropping the level of the ocean and exposing land that was once beneath the sea.

CLIMATE CHANGE

During the last ice age, the global climate warmed abruptly several times over the 100,000-year interval. Each new cycle included a buildup of the

North American ice sheet whose central region was 2 miles or more thick. The series of warming fluctuations in the world's temperature were followed by the breakup of the ice sheet into armadas of icebergs that invaded the North Atlantic (Fig. 5–1). Six distinct iceberg flotillas occurred around 65,000, 52,000, 39,000, 23,000, and 16,000 years ago.

At the peak of the ice age, icebergs covered up to half the area of the oceans. Their high albedo reflected sunlight back into space, which maintained cool global temperatures and allowed the glaciers to continue growing. Large numbers of icebergs calving off the glaciers entering the sea also acted like giant ice cubes, significantly lowering ocean surface temperatures (Fig. 5–2), which curtailed ocean circulation systems.

The lowered temperatures reduced the evaporation of seawater from the ocean, dropping precipitation rates over the continents. Because little melting occurred during the cooler summers, less snowfall was needed to

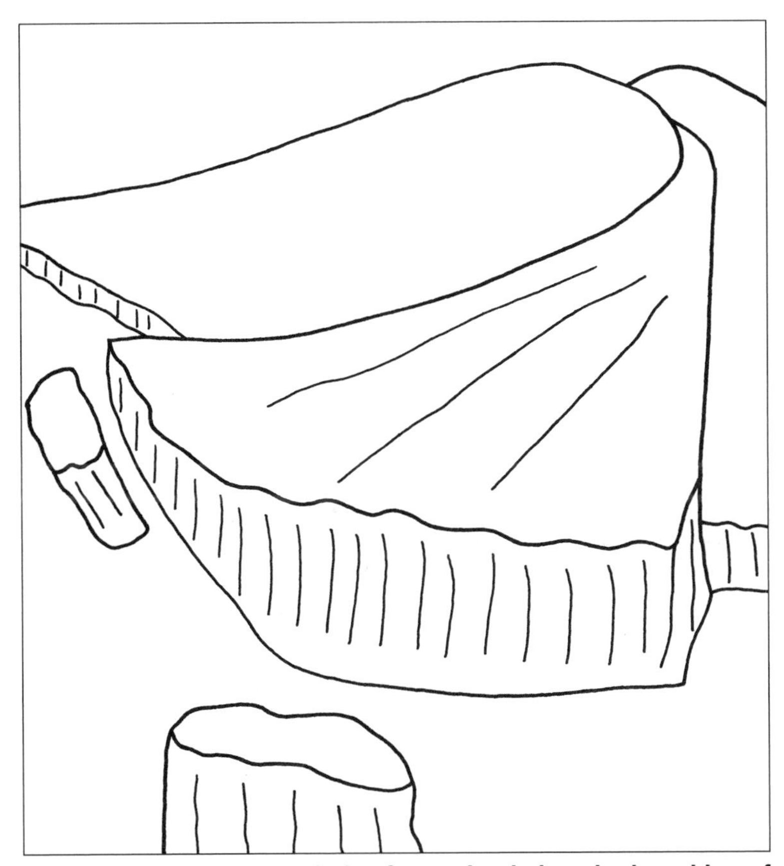

Figure 5–1 Recurrent periods of warming led to the launching of great armadas of icebergs from the North American ice sheet into the North Atlantic Ocean during the last ice age.

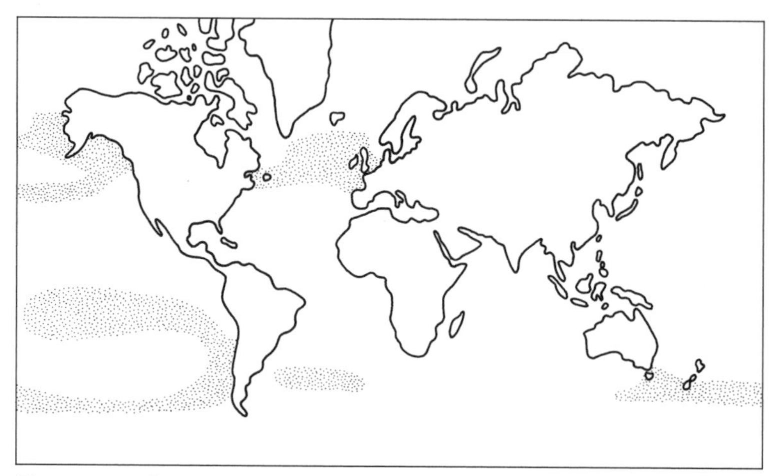

Figure 5–2 Regions of August sea-surface cooling during the last ice age as shown by the stippled areas.

maintain the ice sheets. The drop in precipitation levels also increased the spread of deserts over the world. Fierce desert winds produced gigantic dust storms, and so much dust clogged the atmosphere it shaded the Earth and kept it cool. Dust landing on the ice sheets significantly reduced their albedo, which might have helped trigger deglaciation.

The glaciers in North America substantially changed atmospheric circulation and altered storm patterns over the continent. Cold winds blowing off the ice sheets also affected the climate of the glacial margins. High-pressure centers over the Laurentide ice sheet brought strong easterly winds along the southern flank of the glacier and strengthened the jet stream aloft (Fig. 5–3). The change in climate produced abundant rainfall in the southwestern deserts of the United States, where woodlands grew soon after the ice sheets began to retreat.

Continents located in the polar regions cause extended periods of glaciation because the land is more reflective of solar radiation and has less heat capacity than the high-latitude seas. Oceans located in the tropics have a net cooling effect because they reflect more solar rays back into space than the land. Moreover, the increased land area in the high latitudes, where snow falls steadily with little melting, establishes a permanent polar glacial climate, which encourages the accumulation of snow and ice.

Once the glaciers are in place, their high reflectivity perpetuates them and sustains glaciation even if the once high land sank to the level of the sea due to a decrease in crustal buoyancy by the heavy overlying layers of ice. The increased weight of ice might actually squeeze magma out of the Earth, causing a rise in volcanic activity. Increasing the weight on the land also could interfere with the flow of mantle currents, possibly causing a rise in volcanism.

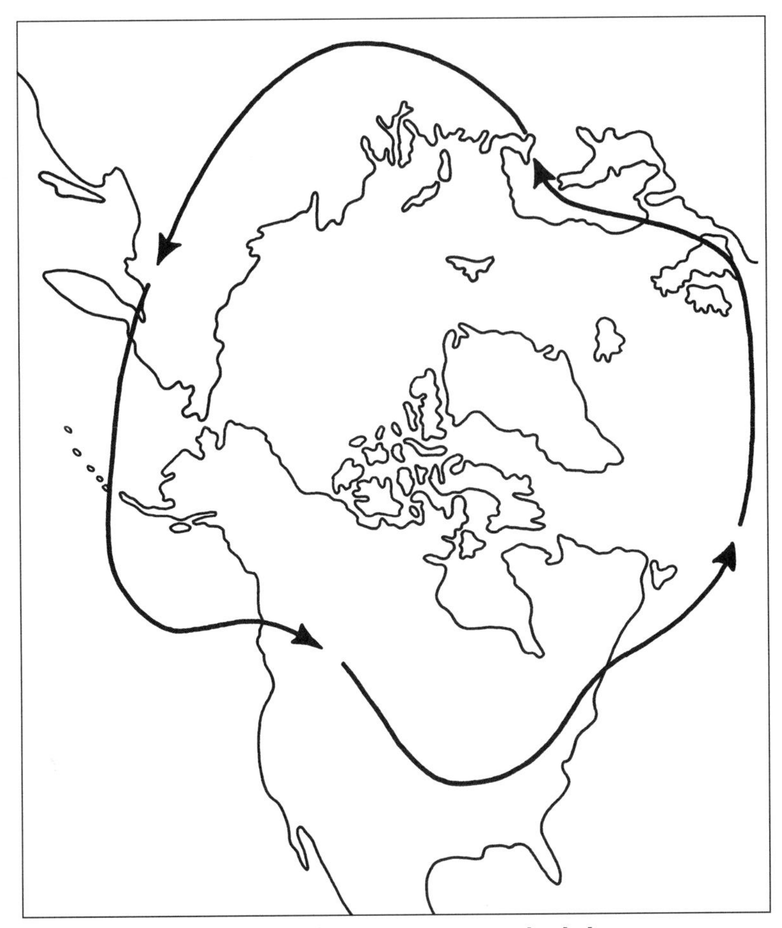

Figure 5–3 Typical flow of the upper atmospheric jet stream.

The greater volcanic activity injects huge amounts of ash into the atmosphere and shades the planet. The lack of sunlight further cools the Earth and increases glaciation, providing an efficient feedback mechanism. With such a system in place, glaciers would seem to be extremely difficult to halt. Yet the geologic record shows that dramatic changes in the climate did occur, forcing the retreat of the ice sheets.

Polar ice drives the atmospheric and oceanic circulation systems (Fig. 5–4), which have a tremendous effect on the global climate. The moisture-laden hot air rising from the tropics is forced to move poleward by cooler, heavier air masses, indirectly cooled by the polar ice. The polar regions, therefore, play a significant role in major long-range alterations in global climatic patterns and have a considerable impact on the weather of the midlatitudes (Fig. 5–5).

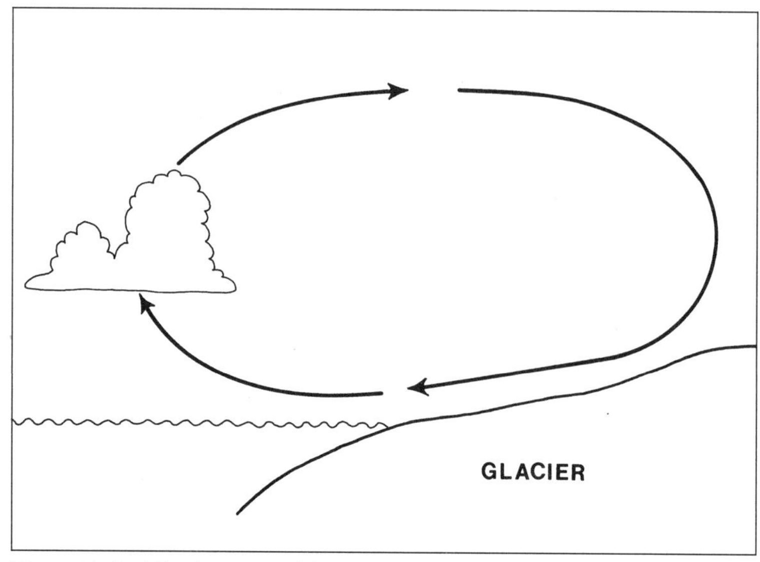

Figure 5–4 The ice caps drive the atmospheric and oceanic circulation.

The fluctuation of the ice cover from year to year dramatically affects the weather. The reduction of the ice cover increases the amount of open ocean and subsequent evaporation, increasing cloud cover. The additional cloudiness decreases temperatures and allows the ice cover to expand, reducing evaporation and cloud cover. Therefore, the ice caps provide an effective means for moderating the global temperature.

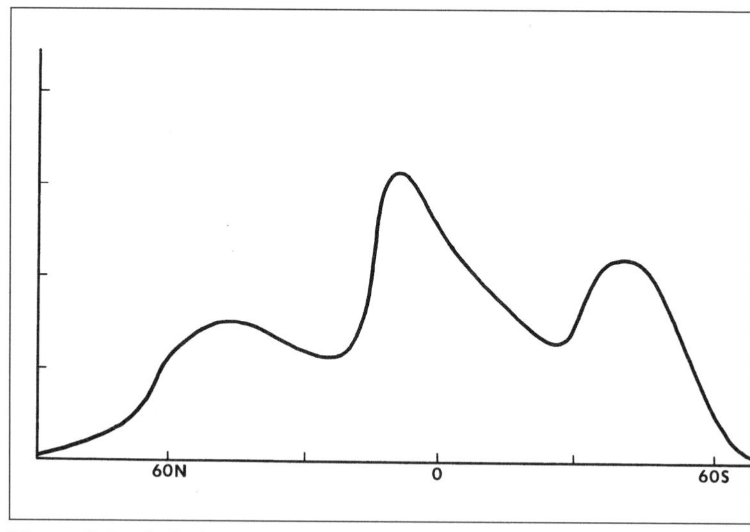

Figure 5–5 Relative precipitation by latitude.

SPECIES DIVERSITY

The ice ages not only dramatically influenced life on this planet for the last 2 billion years or so, but life itself might have significantly changed the climate to initiate glaciation. Early in the Earth's history, photosynthetic organisms began substituting oxygen for carbon dioxide in the atmosphere, which weakened the greenhouse effect and lowered global temperatures. Life clearly influences the composition of the atmosphere and oceans, and without it the climate system would be totally out of control.

The most important factor affecting the diversity of species is climate cooling, and the greatest factor limiting the geographic distribution of marine species is ocean temperature. Therefore, the cooling of the climate

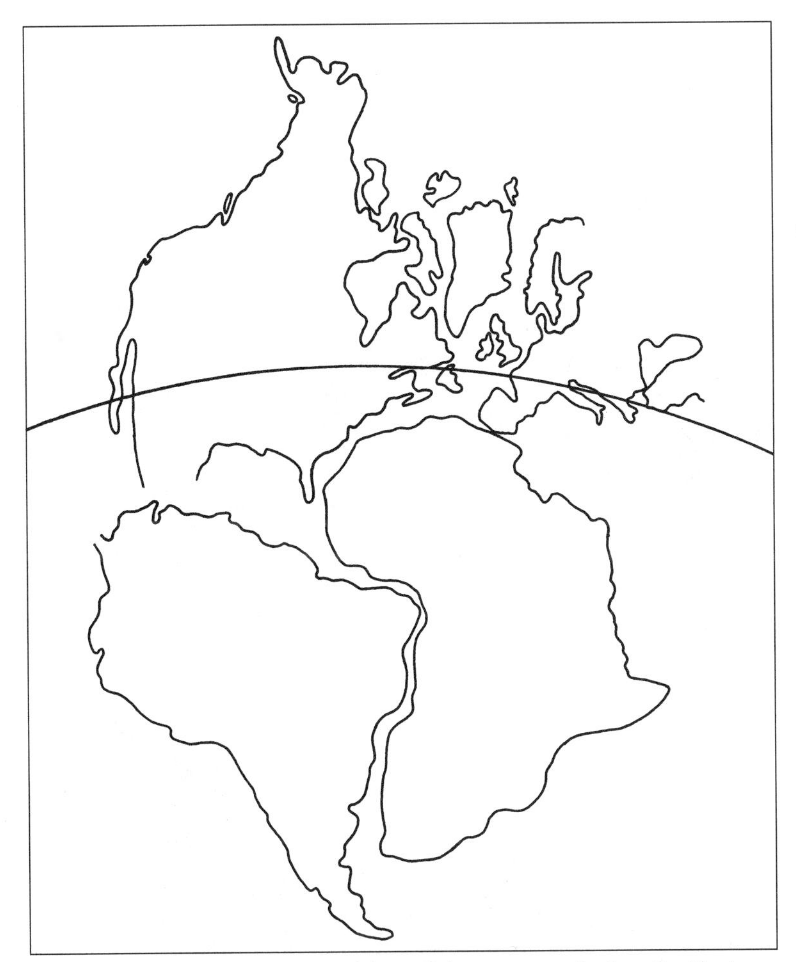

Figure 5–6 Approximate position of the equator during the Carboniferous period, when much of the Earth was covered by dense forests.

is primarily responsible for most crises among seafaring creatures. As the world's oceans cool, mobile species are forced to migrate into the warmer regions of the tropics.

Climate cooling is the primary culprit behind most extinctions in the sea. Species unable to migrate to warmer regions or adapt to colder conditions are usually the hardest hit, especially tropical faunas that only tolerate a narrow range of temperatures and are unable to migrate. Since lowered temperatures also slow down the rate of chemical reactions, biological activity during a major glacial event is expected to function at a lower energy state as well, which also affects the rate of evolutionary growth and species diversity. Only those species that have previously adapted to cold conditions still thrive in today's oceans. They are mostly herbivores that tend to be generalized feeders, consuming many types of vegetation.

The geographic positions of the continents with respect to each other and to the equator (Fig. 5–6) helped determine climatic conditions. When most of the land resided in the tropics, the climate was warm. But when lands wandered into the polar regions, the climate turned cold, spawning episodes of glaciation and mass extinction. Continental movements alter the shapes of ocean basins, which affects the flow of ocean currents, the width of continental margins, and the abundance of marine habitats.

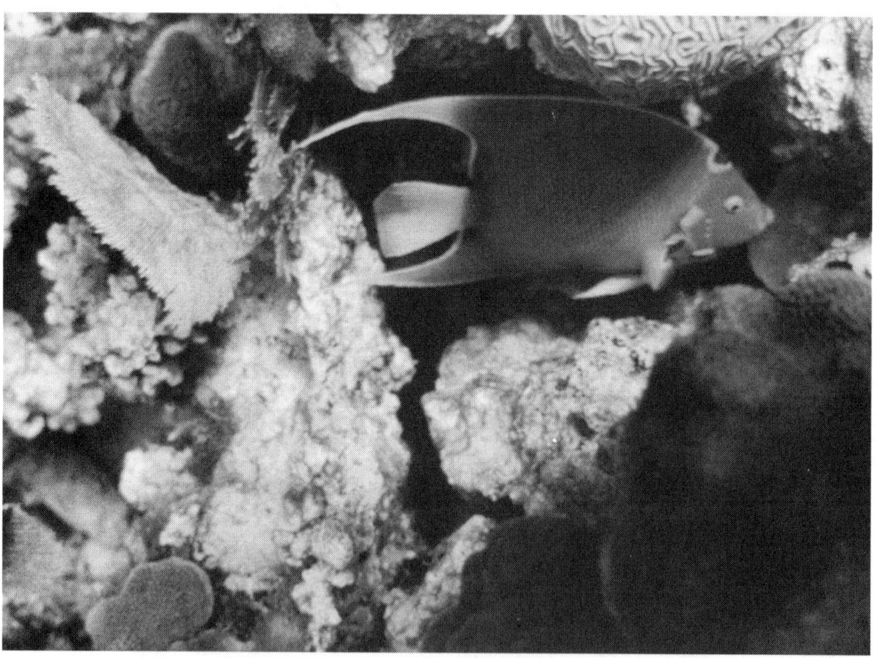

Figure 5–7 An angelfish swims among rock and coral near High Bay, Andros Island, Bahama Islands. Photo by P. Whitmore, courtesy of U.S. Navy

These cooling events removed the most vulnerable of species, so that those living today are more robust and can withstand the extreme environmental swings during the last 3 million years, when glaciers spanned much of the Northern Hemisphere. This might explain why no major extinction occurred during the Pleistocene, unlike other glacial periods. The oldest species living in the world's oceans today thrive in cold waters. Many Arctic species, including certain brachiopods, starfish, and bivalves, belong to biological orders whose origins extend hundreds of millions of years into the Paleozoic era. In contrast, tropical faunas such as reef communities (Fig. 5–7) were constantly battered by periodic mass extinctions.

GLACIAL EXTINCTION

Many episodes of extinction coincided with periods of glaciation, and the effect of global cooling on life is considerable. Furthermore, changes in seawater chemistry due to additional sea ice are often invoked to explain biological extinctions in the ocean. All known episodes of glaciation are associated with lower sea levels, although not all mass extinctions correlated with the lowering of the sea.

Mass extinctions generally accompany periods of glaciation because the living space of warmth-loving species is restricted to areas around the tropics. Species trapped in confined waterways and unable to move to warmer seas were particularly hard hit. Furthermore, the accumulation of glacial ice in the polar regions lowered sea levels, thereby reducing shallow water shelf areas, which limited the habitat area and consequently the number of species it could support.

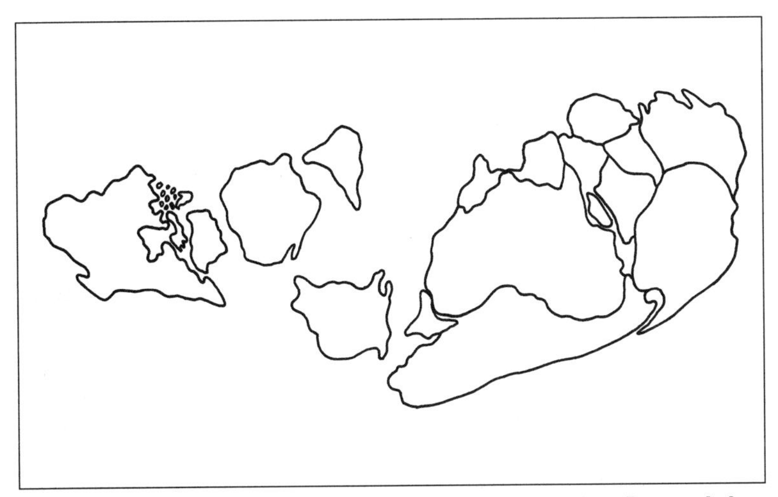

Figure 5–8 Distribution of the continents during the late Precambrian.

Figure 5–9 Fossil brachiopods and trilobites from the Bonanza King Formation, Trail Canyon, Death Valley National Monument, Inyo County, California. Photo by C. B. Hunt, courtesy of USGS

During the late Precambrian, when life was sparse, several ice ages occurred and great ice sheets covered more than half the land surface, instigating the first major extinction around 670 million years ago. All continents assembled into a supercontinent that might have wandered over one of the poles (Fig. 5–8). The extinction decimated the ocean's population of single-celled phytoplankton, which were the highest form of life at that time and the first organisms with cells containing nuclei. When the glaciation ended, a rapid population growth ensued, with a diversification of species having no equal in Earth's history.

At the end of the Ordovician period, around 440 million years ago, a mass extinction eliminated some 100 families of marine animals. Glaciation reached its peak as ice sheets radiated outward from a glacial center in North Africa, which hovered directly over the South Pole. Most of the victims were tropical species sensitive to large fluctuations in the environment. Among those that became extinct were the graptolites, which were colonies of cup-shaped organisms that resembled a conglomeration of floating stems and leaves, along with many species of trilobites (Fig. 5–9), oval-shaped arthropods and ancestors of today's horseshoe crab.

Many tropical marine groups disappeared during a major extinction event near the end of the Devonian period about 365 million years ago, possibly due to climate cooling. Much of Gondwana was in the southern polar region during the Devonian, and seas flooded broad areas of the continent. The extinction apparently occurred over a period of 7 million years and eliminated species of corals and many other bottom-dwelling marine organisms. Primitive corals and sponges, which were prolific limestone reef builders early in the period, suffered heavily during the extinction and never fully recovered.

As these animal groups were vanishing, the glass sponges, which were more adaptable to cold conditions, were generally unaffected and rapidly diversified, only to dwindle when the crisis subsided and other groups recovered. The sponges' prosperity signifies that less fortunate species had succumbed to the effects of climate cooling. Large numbers of brachiopod families also died out at the end of the Devonian. In contrast, cold-adapted animals living in the Arctic waters fared quite well.

In the late Carboniferous, Gondwana drifted into the south polar regions, where glacial centers expanded across the large continent (Fig. 5–10). Land once covered with great coal swamps dried out as the climate grew colder.

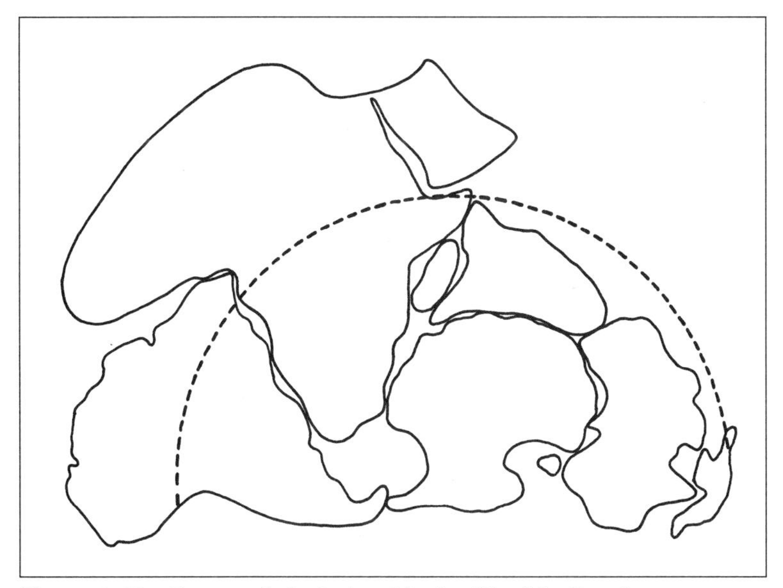

Figure 5–10 Extent of glaciation (dashed line) in Gondwana.

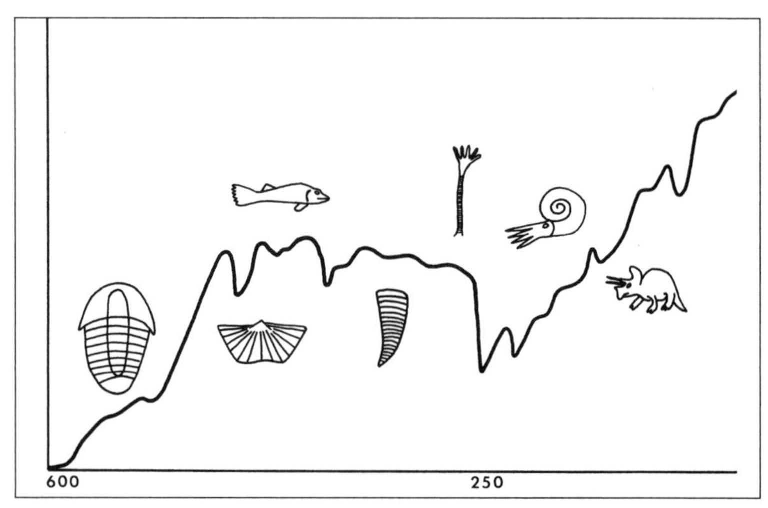

Figure 5–11 The diversification of species. The large dip is in response to the great Permian extinction. Dates in millions of years ago.

No major extinction event occurred during the widespread Carboniferous glaciation around 330 million years ago, however. The relatively low extinction rates were probably the result of a limited number of extinction-prone species following the late Devonian extinction.

The greatest extinction event occurred at the end of the Permian period about 250 million years ago (Fig. 5–11). It was particularly devastating to

TABLE 5–1 FLOOD BASALT VOLCANISM AND MASS EXTINCTIONS

Volcanic episode	Million years ago	Extinct event	Million years ago
Columbian River, U.S.	17	Low-mid Miocene	14
Ethiopian	35	Upper Eocene	36
Deccan, India	65	Maastrichtian	65
		Cenomanian	91
Rajmahal, India	110	Aptian	110
South-West African	135	Tithonian	137
Antarctica	170	Bajocian	173
South African	190	Pliensbachian	191
East North American	200	Rhaectian/Norian	211
Siberian	250	Guadalupian	249

marine fauna. Half the families of marine organisms, including more than 95 percent of all known species, abruptly disappeared. Planktonic ocean plants died in a geologic instant, rapidly knocking out the bottom of the marine food chain. The extinction followed on the heels of the late Permian glaciation, when a thick sheet of ice blanketed much of the Earth, significantly lowering ocean temperatures. The extinction also coincided with massive outpours of lava in Siberia, the largest basalt flows known on Earth

During the last 100 million years, extinction events generally coincided with three major steps in the evolution of the global climate. The first was

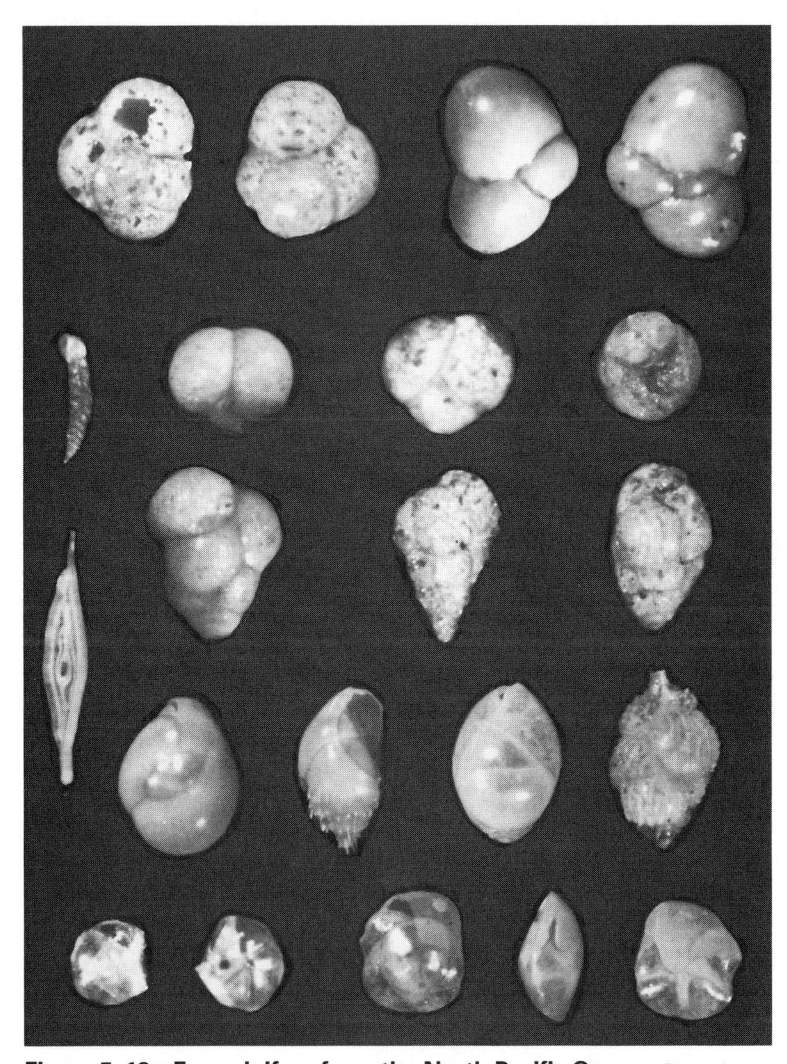

Figure 5–12 Foraminifera from the North Pacific Ocean. Photo by P. B. Smith, courtesy of USGS

the onset of the midlatitude Northern Hemispheric glaciation from 2.4 to 3.0 million years ago. The second was a major expansion of ice on Antarctica between 10 and 14 million years ago. And the third was a major cooling event between 31 and 40 million years ago, when Antarctica acquired its first and largest ice sheet. Another mass extinction coincided with a major environmental change about 90 million years ago.

Near the end of the Eocene epoch about 37 million years ago, global temperatures plummeted. Antarctica separated from Australia, wandered over the South Pole, and was buried under a thick mantle of ice. Glaciers also grew for the first time in the highest ramparts of the Rocky Mountains, which rose during the Laramide orogeny, a mountain-building episode from 80 million to 40 million years ago. Moreover, the Himalaya Mountains and the Tibetan Plateau, (which itself soars above 3 miles), were uplifted during this time when India collided with southern Asia.

The period also coincided with mass extinctions in the ocean, eliminating many types of plankton, and on the land, which saw the abrupt disappearance of the archaic mammals. Afterward, the descendants of modern mammals evolved in their place. Major changes in deep-ocean circulation also correlated with the late Eocene extinctions. The extinctions eliminated many European species of marine life, when shallow seas flooded the continent. The separation of Greenland from Europe during this time might have been responsible for the draining of frigid Arctic waters into the Atlantic, significantly lowering the ocean's temperature and causing the disappearance of most types of foraminifera (Fig. 5–12).

POLAR WANDERING

Over the past 4 centuries, navigational charts have revealed two major trends in variations in the geomagnetic field (Fig. 5–13). The first is a slow, steady decrease in the intensity of the field at such a rate that it could collapse altogether. Apparently, the geomagnetic field remains stable for long periods. Then for unknown reasons, the electrical currents in the Earth's core, which generate a self-inducing dynamo effect, fail and the magnetic field collapses, eventually regenerating with opposite polarity. The second variation is a slow westerly drift in irregular eddies in the field, amounting to about 1 degree of longitude every 5 years. The drift suggests that the fluid in the outer metallic core, which generates the geomagnetic field, is moving at a rate of about 300 feet a day.

Due to the Earth's rotation, the magnetic field aligns in one direction. The magnetic poles do not coincide with the axis of rotation but are presently offset at an angle of about 11 degrees. The poles are not stationary, either, but slowly wander around the polar regions. Since the turn of the century, the North Pole has crept about 30 feet toward eastern Canada. At this rate

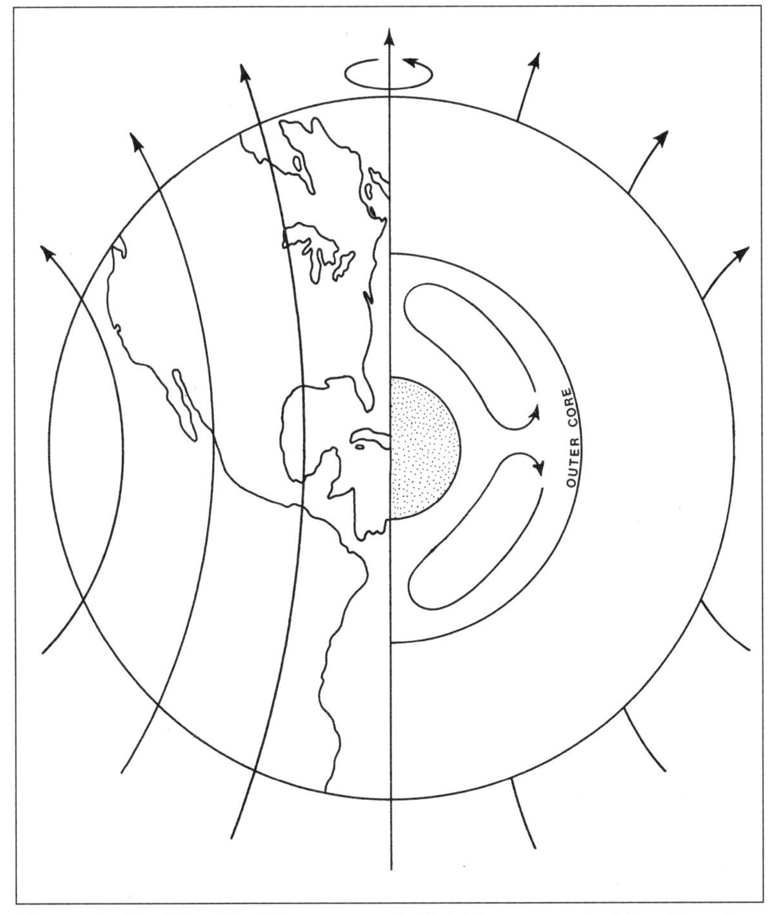

Figure 5–13 The Earth's magnetic field is generated by the slow rotation of the solid metallic inner core with respect to the liquid outer core.

of polar wandering, in 10 million years, Philadelphia could be 10 degrees closer to the North Pole.

Polar wandering alters the direction of the Earth's magnetic field, and these changes are recorded in the rocks. Therefore, rocks formed at different times are imprinted with different inclinations. The paleomagnetic evidence shows that the North Pole wandered some 13,000 miles over the last billion years from western North America, across the northern Pacific Ocean and northern Asia, finally coming to rest at its present location in the Arctic.

The great ice sheets that have advanced and retreated over the last 3 million years might have been massive enough to cause the Earth to reorient itself and produce the present polar wandering (Fig. 5–14). This is because

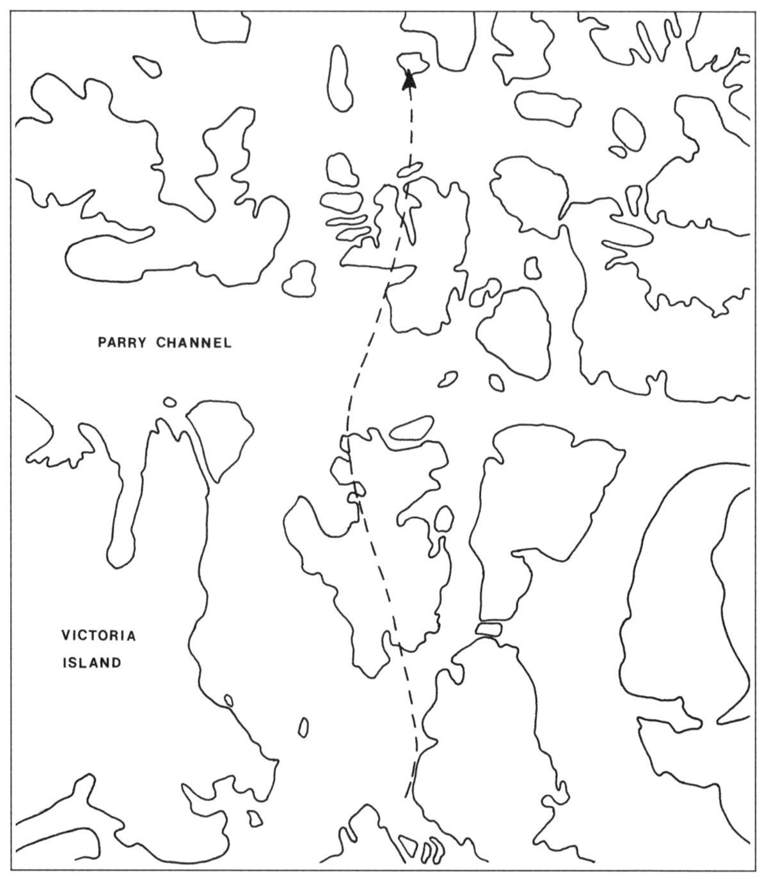

Figure 5–14 About 150 years of the North Magnetic Pole wandering from the Boothia Peninsula to Ellef Ringness Island in the Canadian Arctic, a distance of approximately 450 miles.

the Earth tends to place most of its mass near the equator to maintain stability, and large ice sheets are massive enough to upset the balance. Furthermore, episodes of polar wandering might be linked with the tendency of the magnetic field to reverse itself.

Two or three times every million years, the Earth's geomagnetic field reverses polarity, with the north and south magnetic poles trading places. In eastern Ireland, some 900 separate lava flows dating to 20 million years ago had 60 complete changes of magnetic polarity, or on average about 3 every million years. Over the last 4 million years, the field reversed 11 times, with the last reversal occurring about 780,000 years ago, indicating that perhaps the next one is well overdue.

Magnetic field reversals correspond to periods of glaciation, and an occurrence around 2 million years ago might have aided the progress of a succession of ice ages. Reversals in the magnetic field and excursions of the magnetic poles appear to correlate with periods of rapid cooling and the extinction of species. The Gothenburg geomagnetic excursion, occurring about 13,500 years ago in the midst of a longer period of rapid global warming, caused temperatures to fall and glaciers to advance for 1,000 years, apparently in response to a weakened magnetic field.

Although changes in the magnetic field do sometimes agree with changes in global ice volume, short-term rapid glaciations, and climate cooling, no decisive evidence has shown that the relationship was anything but coincidental. Yet internal turmoil in the core, which upsets the magnetic field and causes it to reverse itself, could be the result of excess heat that produces erupting plumes of hot rock in the mantle.

When a plume passes the boundary between the lower and upper mantle 600 miles below the surface, the bulbous head separates and rises toward the surface, followed millions of years later by a newly created plume. This translates on the Earth's surface as an increase in seafloor spreading rate due to a larger amount of volcanism on the ocean floor and on the continents.

During periods of intense plume activity, almost no magnetic reversals occur. Only when the plume activity is low, do magnetic reversals take place at a high rate. After a period of several hundred thousand to a million years or more of stability, when the number of thermal plumes increased inside the mantle, the strength of the magnetic field gradually decays over a short time span, probably no more than 10,000 years. It then abruptly collapses, reverses, and slowly builds back to its normal strength. The

TABLE 5–2 COMPARISON OF MAGNETIC REVERSALS WITH OTHER PHENOMENA (DATES IN MILLIONS OF YEARS)

Magnetic reversal	Unusual cold	Meteorite activity	Sea level drops	Mass extinctions
0.7	0.7	0.7		
1.9	1.9	1.9		
2.0	2.0			
10				11
40			37–20	37
70			70–60	65
130			132–125	137
160			165–140	173

process might take upward of 1,000 years before the magnetic field regains its full intensity.

An apparent correlation exists between changes in the magnetic field and other phenomena taking place on the Earth's surface (Table 5-2). A comparison of the reversals with known variations in the climate shows a striking agreement in many respects. Magnetic reversals occurring 2.0, 1.9, and 0.7 million years ago coincided with unusual cold spells as suggested by the ratio of carbon and nitrogen in ancient lake bed sediments. High ratios indicate dwindling nitrogen levels due to shrinking populations of algae and plankton in response to colder climates.

OCEAN CIRCULATION

Surface and abyssal currents play a vital role in moving heat around the planet (Fig. 5–15). Steady winds blowing across the ocean drive the surface currents, and, as with flowing air masses, the Coriolis effect caused by the Earth's rotation deflects them to the right in the Northern Hemisphere and to the left in the Southern Hemisphere. Surface currents in the ocean are similar in function to air currents. They transport warm water from the tropics, distribute it to the higher latitudes, and return with colder water, only they travel much more slowly.

Abyssal currents are driven by thermal forces. The sinking of cold, dense water near the poles generates strong, deep currents flowing steadily

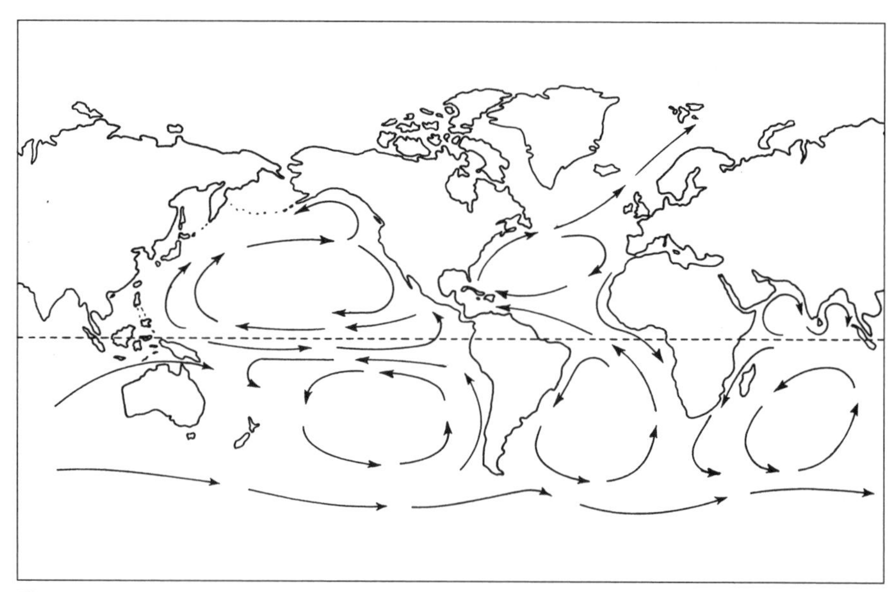

Figure 5–15 The major ocean currents distribute the Earth's heat around the globe.

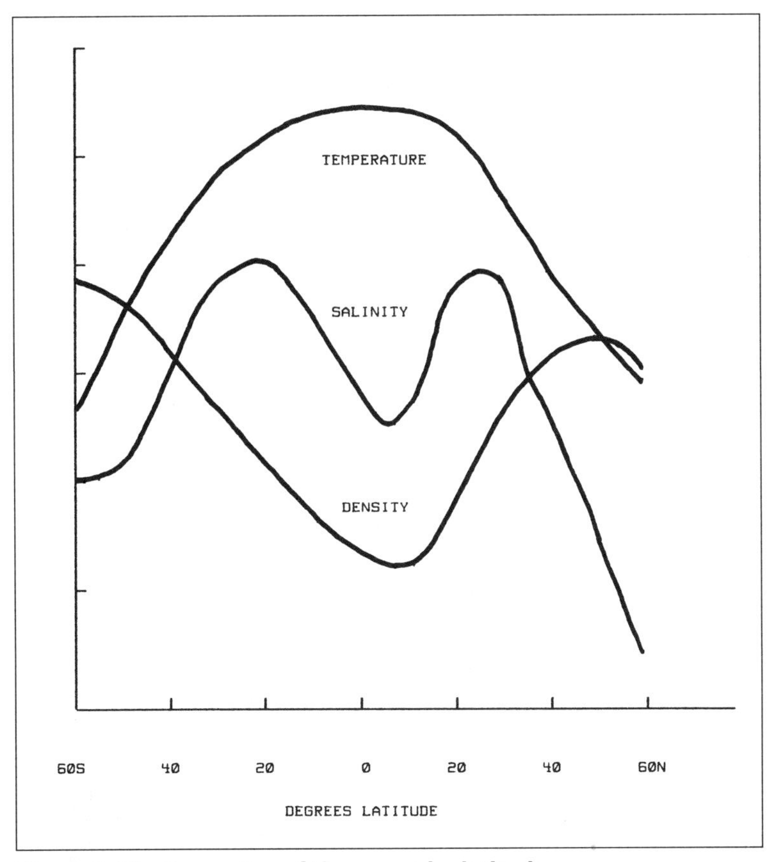

Figure 5–16 Properties of the ocean by latitude.

toward the equator. Like surface currents, abyssal currents deflect westward because of the Earth's eastward rotation. The distribution of landmasses and the topography of the ocean floor, including ridges and canyons, also control the path taken by the circulating water.

Vertical mixing between ocean layers occurs very slowly because water density increases with depth, inhibiting mixing of different layers. Rather than crossing the density surfaces, water moves far more readily along the plane of a single layer. These density surfaces are inclined instead of horizontal due to the forces induced by the Earth's rotation. Therefore, water rises and sinks by moving along one of the sloping density layers.

The ocean is mostly filled with icy cold water only a few degrees above zero that was chilled while at the surface of the polar seas. The Indian Ocean is unique because it is not in contact with the north polar region and has only one source of cold bottom water from the Antarctic. By contrast, the Atlantic and Pacific oceans connect with both the Antarctic and the Arctic oceans.

The surface water in the polar regions is denser than in other parts of the world due to its low temperature and high salt content. Water attains its highest density at 4 degrees Celsius. The cold, dense seawater forms at high latitudes by heat loss to the atmosphere. The increased saltiness results from the evaporation of poleward flowing water and the exclusion of salt from ice as it freezes. As the cold seawater increases density, it sinks to the bottom, spreads out upon hitting the ocean floor, and heads toward the equator, where upwelling currents bring the cold water back to the surface and distribute it in the tropics.

Changes in ocean water salinity affect the formation of sea ice (Fig. 5–16). Lowered amounts of freshwater entering the Arctic Ocean increases its salinity. The saltier, heavier water subsequently sinks to the bottom,

Figure 5–17 Grounded growlers in the Arctic Ocean. Photo by Katz, courtesy of U.S. Coast Guard

causing the ocean to overturn. This forces colder, less salty water to the surface, where it freezes into sea ice (Fig. 5–17). The amount of sea ice in the Arctic Ocean varies from over 3 million square miles in the summer to nearly twice that amount in the winter. If none of the ice melted during the short warm season, it could continue to expand onto the continents.

The Antarctic plays a larger role in global ocean circulation than the Arctic. Deep, cold currents flowing from Antarctica toward the equator trend to the left and press against the western side of the Atlantic, Pacific, and Indian ocean basins. As they sweep against the continents, the currents pick up speed similar to the way a stream flows faster in a narrowing channel.

The Atlantic Bottom Water, which is the largest mass of bottom water in the world, is a deep current that travels 7,500 miles from its source in the Antarctic and sweeps along the edge of the abyssal plain south of Nova Scotia. It sinks from the surface near Antarctica and flows northward along the seafloor into the western North Atlantic. Before mixing with North Atlantic water and dispersing, some of this flow curves to the west due to the Earth's eastward rotation and hugs the lower edge of the continental rise at the border of the abyssal plain.

The North Atlantic Deep Water, which flows south from Greenland and around the tip of Africa, plays an important role in keeping the climate of western Europe fairly mild. The current forms when dense, salty water from the south cools near Greenland and grows heavy, causing it to sink to the ocean bottom. The deep-flowing current turned on abruptly around 12,500 years ago, when the ice sheets began rapidly melting, which might have initiated the shifts in ocean circulation. Otherwise, some previous melting or an increase in greenhouse gases might have altered circulation patterns in the ocean.

Massive amounts of ice calving off the retreating glaciers at the end of the last ice age apparently crashed into the North Atlantic, disrupting deep-water formation and the transfer of heat northward. The melting ice sheets sent a torrent of meltwater and icebergs spilling into the North Atlantic and rerouted ocean circulation, initiating a return to ice age conditions. The massive floods formed a cold, freshwater lid on the ocean that significantly changed the salinity of the seawater. The cold waters also blocked poleward flowing warm currents from the tropics, causing land temperatures to fall to near ice age levels.

A cold spell hit northern Europe, Greenland, and the Atlantic coast of Canada, sending these regions back into near glacial conditions. The system of ocean currents that warms Europe ceased operating, and sea ice covered much of the North Atlantic during this period. When the cold spell ended, temperatures rapidly climbed to present interglacial conditions within a few decades. A renewal of the deep-ocean circulation system, which shut off entirely or became severely weakened during the ice age, might have thawed out the planet.

Major changes in the ocean's circulation appeared to have triggered the cold spells of the previous interglacial as well. Periodic halts in the northward flow that carries warm Atlantic water to the far northern North Atlantic are a plausible cause of the climatic oscillations around the end of the last ice age. The shutdowns in the ocean current might have stemmed from occasional collapses of the ice sheets into the ocean, which also would raise the level of the sea.

SEA LEVEL CHANGES

One way to rapidly lower the sea is for massive glaciers to grow in the higher latitudes. They would, in turn, substantially affect the climate by altering atmospheric and oceanic circulation patterns. However, no geologic evidence exists for polar ice of any significance until the beginning of the Oligocene epoch about 37 million years ago, when Antarctica was buried under a massive ice sheet. Perhaps, instead, the land rose higher due to an increase in tectonic activity that uplifted extensive portions of the continents, causing a regression of the sea.

In the Oligocene, seas that overrode the continents receded as the ocean withdrew to one of its lowest levels over the past several hundred million years, possibly due in part to the accumulation of massive ice sheets on Antarctica. Although the sea level remained depressed for 5 million years, little or no excess extinction of marine life occurred. Therefore, crowding conditions brought on by lowering sea levels cannot be responsible for all the extinctions. Moreover, during many mass extinctions, the sea level was not much lower than it is today.

During the last ice age, glacial ice locked up about 5 percent of the Earth's water. So much water was removed from the ocean that it dropped as much as 400 feet lower than at present. This exposed several land bridges and aided the migration of animals from one continent to another. It also allowed Native Americans to cross over into North America from Asia via the Bering Strait, possibly as early as 30,000 years ago.

The melting of the great ice sheets at the end of the last ice age between 14,000 and 7,000 years ago, sent floods of meltwater into the ocean, raising the sea level on average perhaps 3 feet per century. In the previous interglacial period, which ended about 115,000 years ago, the climate was actually warmer than it is today, and the melting of the ice caps caused the sea level to rise about 60 feet higher than today.

All known episodes of glaciation were accompanied by lowered sea levels. The accumulated ice during the last glaciation dropped the level of the ocean sufficiently to advance the shoreline seaward up to 100 miles or more (Fig. 5–18). The coastline of the eastern seaboard of the United States extended about halfway to the edge of the continental shelf, which stretches more than 600 miles eastward. The drop in sea level exposed land bridges

Figure 5–18 Extended shoreline during the height of the last ice age.

and linked continents, allowing species to island hop from one part of the world to another.

The falling sea levels also replenish sand on beaches along the east coast of North America. Most of the sand on the coast and continental shelf originated in the north from sources such as the Hudson River. In order for the sand to move as far south as the Carolina coast, it had to progress in steps that possibly took millions of years as sea levels rose and fell in concert with the ice ages.

As the sand moves along a coast, ocean currents push it into large bays or estuaries. The embayment continues to fill with sand until sea levels drop and the accumulated sediment flushes down onto the continental shelf. The sand can travel only as far as the next bay in a single glacial cycle, however. Therefore, most beaches will not be restocked with sand until the next ice age.

6

CONTINENTAL GLACIERS

Other than during the last few million years, our planet appears not to have had permanent ice caps in both polar regions. Even a single ice sheet at the poles was a rare event in geologic history. The formation of polar ice depends mostly on the location of the surrounding landmasses. Throughout most of the Earth's history, continents were confined to regions around the equator, allowing warm ocean currents access to the poles, which kept them icefree year-round. Within the last 50 million years, however, the continents shifted positions so that a large continent overlies the South Pole and a nearly landlocked sea covers the North Pole.

Most of the continental landmass also moved north of the equator, leaving little land in the Southern Hemisphere, which is mostly ocean. The drifting of the continents radically changed the pattern of ocean currents, whose access to the poles was severely restricted. Without warm ocean currents from the tropics to keep the polar regions free of ice, glaciers will remain until the continents again move back toward the equator.

THE POLAR ICE CAPS

About 3 percent of the planet's water is locked up in the polar ice caps, which cover on average about 7 percent of the Earth's surface. Excluding ice in the polar regions, the remaining glaciers on the continents hold as much freshwater as all the world's rivers and lakes. The Arctic is a sea of ice (Fig. 6–1a and b), whose boundary is the 10 degree Celsius July isotherm, which is the extent of the polar pack ice during summer. The medium extent of the Arctic ice cap is approximately 65 degrees north latitude. Sea ice covers an average area of about 4 million square miles, with an average thickness of 15 to 20 feet.

During the Antarctic winter, from June to September, sea ice covers nearly 8 million square miles of ocean surrounding the continent, with an average thickness of about 10 feet. Unlike the Arctic ice pack, which is relatively stable and well understood, the Antarctic ice pack is neither of these. Ice forms only during the winter months in most of the waters off Antarctica, and the amount varies from year to year.

Apparently, no permanent ice capped either pole prior to the Cenozoic era, beginning 65 million years ago. The deep ocean waters, which are now near freezing, were around 15 degrees Celsius (60 degrees Fahrenheit) during most of the preceding Mesozoic era. The average global surface temperature, which today is about 15 degrees Celsius, was 10 to 15 degrees warmer. The polar regions were much warmer than at present, with no evidence for any permanent ice caps. The temperature difference between the poles and the equator was only about 20 degrees, whereas today it is nearly doubled.

Continental glaciers are the largest ice sheets. During periods of glaciation, they covered as much as one-third of the land surface. Presently, only

Figure 6–1a The nuclear-powered submarine USS *Queenfish* at the North Pole.
Photo by C. L. Wright, courtesy of U.S. Navy

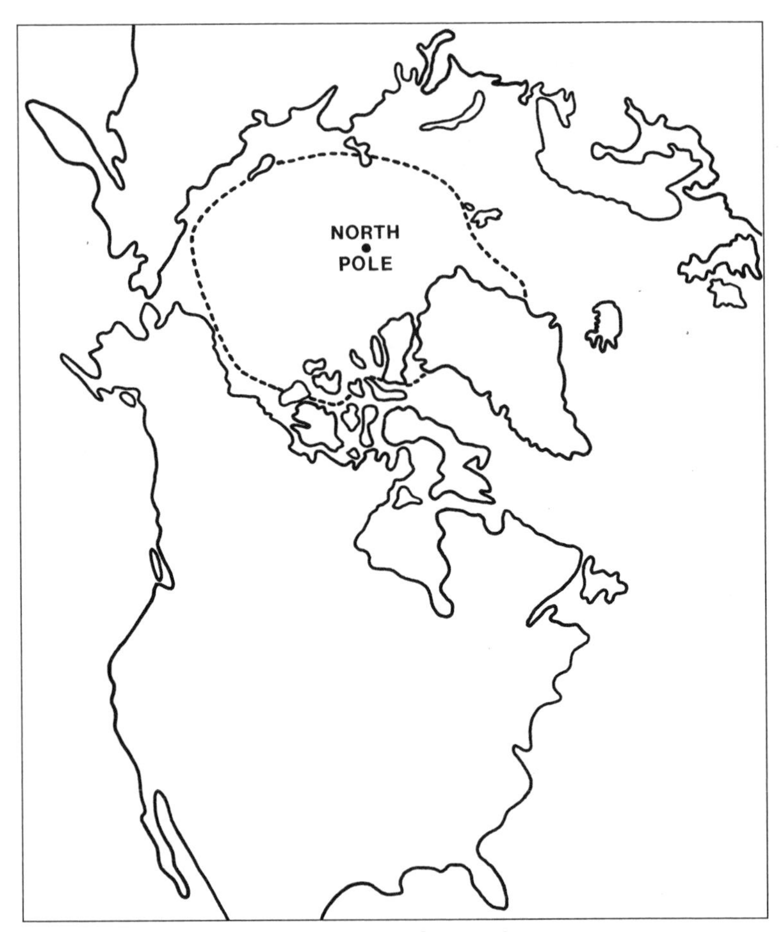

Figure 6–1b Extent of Arctic ice in September.

Antarctica and Greenland have substantial ice masses, containing only about one-third of the ice cover that existed during the last ice age. A continental glacier moves in all directions away from its point of origin and completely engulfs the land except for isolated mountain peaks, which project above its surface. The term ice cap also describes a small glacier that spreads out radially from a center such as the ice sheet on Iceland, which is only a minor patch of ice compared to its ice age counterpart.

A glacier begins when winter snowfall exceeds summer snowmelt and accumulates into a snowfield. Most of the glacial ice originates as fresh, light snowflakes, accumulating above the snow line for alpine glaciers (Fig. 6–2) and north of the Arctic and Antarctic circles for continental glaciers. The weight of the snow compresses the lower layers along with a downward seeping and freezing of surface water. The snow crystals are compressed

Figure 6–2 White Chuck Glacier at Glacier Peak Wilderness area, Snohomish County, Washington. Photo by A. Post, courtesy of USGS

and recrystallize into granular ice pellets called firn or névé, which change into solid ice with further compression.

As the ice thickens, it begins to slowly flow like a viscous solid under the influence of gravity. Ice is remarkably fluid and moves by changing shape comparable to the molding of red-hot steel. It also can move by slipping or shifting along cracks through the ice that are like fault lines. An ice stream flows through stationary or slower-moving ice, whereas a glacier is bounded on each side by rock where it moves through mountain valleys. Glacier flow develops on soft beds when the water-saturated sediment becomes so weak it can no longer withstand the stress forces applied by the overlying ice, and the sediments begin to deform. Where the ice stream breaks away from the less mobile ice, one or more deep, gaping cracks, or crevasses, form (Fig. 6–3).

The layering of the glacier is due to alternate melting and freezing and to the accumulation of dust, volcanic ash, and other substances deposited from the atmosphere. Analysis of Antarctic and Greenland ice cores for these materials helps determine the climatic conditions of the past. Bands interspersed through the ice represent shearing and differential movement within the glacier as well as concentrations of debris in these zones.

The movement of glaciers consists of internal deformation, or creep, and basal sliding. Under the great pressure exerted by the weight of the ice, the ice crystals rearrange into layers parallel to the glacier's surface. The gliding of these layers over each other causes the ice to creep. Basal sliding is the movement of a glacier over bedrock under the pull of gravity. Although glacial movement usually combines both of these basic processes, the contribution of each varies greatly with the local conditions. Polar glaciers exhibit little or no basal sliding, whereas the movement of certain temperate glaciers is almost always due to this process because of the lubricating effect of meltwater beneath them.

The rate of creep and basal sliding depends on the thickness, slope, and the ice temperature. Ice produces heat during freezing and absorbs it while

Figure 6–3 Crevasse in the Matanuska Glacier, Alaska. Photo by J. R. Williams, courtesy of USGS

melting. In addition, the rate of flow varies within sections of an individual glacier. Therefore, the ice at the head and terminus of a glacier might move more slowly than in the center. Moreover, velocity tends to decrease from the surface of the glacier down to bedrock.

The largest continental glacier lies atop Antarctica and covers about 5.5 million square miles, an area larger than the United States, Mexico, and Central America combined. Entire mountain ranges are buried by a sheet of ice 3 miles thick in places, with an average thickness of 1.3 miles. The total amount of Antarctic ice is approximately 7 million cubic miles, enough to make an enormous ice cube nearly 200 miles on each side. The ocean covers an area of about 140 million square miles with an average depth of about 2.3 miles. Therefore, if all the ice in Antarctica melted, it would raise the level of the sea an additional 300 feet.

The ice caps moderate the global temperature. If they suddenly disappeared, the Earth's climate would change significantly due to the breakdown in the circulation system that transfers heat from the tropics to other parts of the world. In effect, the tropics could become too hot for successful habitation. The poles would not stay ice free for long, however, because of the lack of warmth from the tropics. The glaciers would quickly grow beyond their present size and overrun the continents, bringing with them the full force of an ice age.

GREENLAND ICE

The largest and only major ice sheet in the Northern Hemisphere overlies Greenland, which happens to be the world's biggest island. It began separating from Eurasia and North America some 57 million years ago. About 4 million years ago, Greenland acquired a permanent ice sheet that is in places 2 miles or more thick. The ice sheet is nourished by the accumulation of snow precipitated from storm systems traversing across its surface. The glacier's growth is balanced by the loss of ice at its boundary regions. Large icebergs calving off glaciers entering the sea play havoc with shipping in the North Atlantic (Fig. 6–4).

Greenland also contains some of the world's oldest rocks. The Isua Formation in a remote mountainous area in southwest Greenland consists of metamorphosed marine sediments formed some 3.8 billion years ago, indicating the Earth had an ocean by this time. Surprisingly, Greenland and Antarctica are devoid of significant earthquakes, probably because the weight of their massive ice sheets presses down hard on the crust and stabilizes existing faults, thus inhibiting fault slip. The ice sheets do not inhibit volcanic activity, however, and underice eruptions can produce massive floods of meltwater.

Greenland was discovered by the Norse in A.D. 982, following their landing on Iceland a decade earlier. Iceland occupies a unique position in

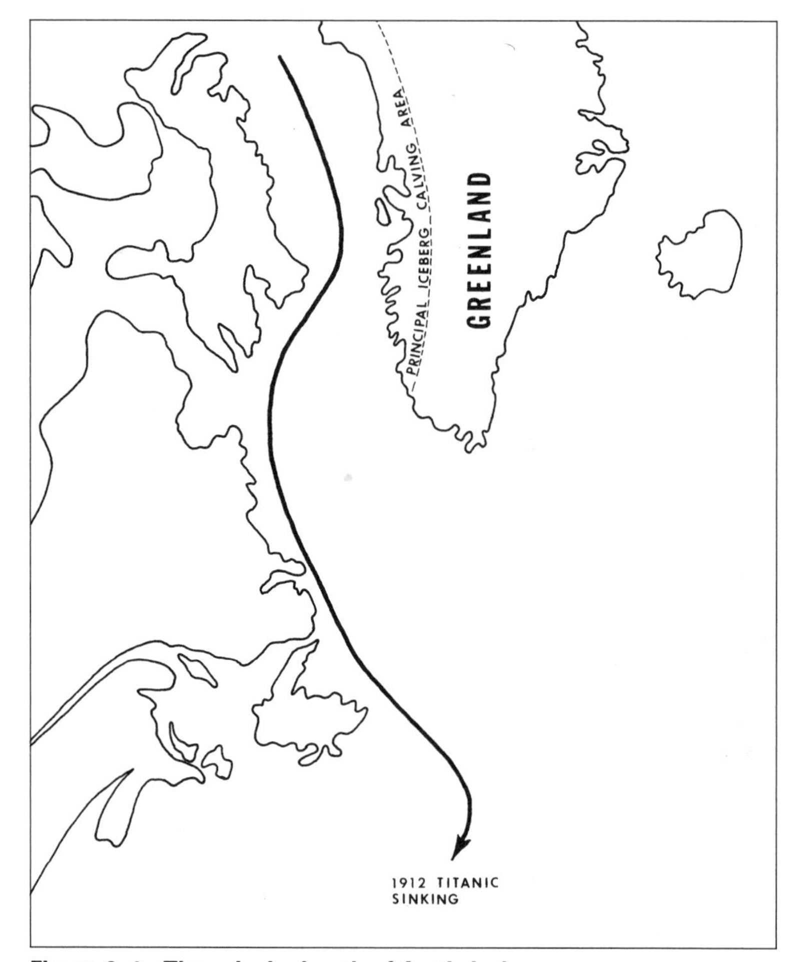

Figure 6–4 The principal path of Arctic icebergs.

the world (Fig. 6–5). As the name suggests, it is one of the coldest inhabited places on Earth and holds the largest ice caps in the Northern Hemisphere outside of Greenland. The Norse, who originally discovered the island, must have been intrigued by the energetic displays of hot water and steam gushing out of the ground, and indeed, the term geyser is from the Nordic word *geysir,* meaning "gusher." The capital city, Reykjavik, which means "smoking bay," was named by the original Norse settlers because of its numerous wisps of steam rising from the surface of the water.

During the Medieval Little Optimum, centered around 1,000 years ago, when temperatures were 1 to 2 degrees Celsius warmer than today, the Greenland ice sheet melted far enough back so that the Norse could settle permanently there. After discovering Greenland, they apparently continued westward and landed in North America, possibly on Labrador, some 500 years before Christopher Columbus discovered the Caribbean. For over 4 centuries the Norse thrived on Greenland, raising cattle and sheep and possibly grain, until the Little Ice Age when advancing glaciers froze them out.

An interesting theory suggests that the frozen island was named Greenland to entice people to settle there. The climate during early colonization was unusually warm so that perhaps parts of the island were green after all. Then around 1450, the climate grew colder, glaciers crept toward the sea, and the Greenland Norse unable to retreat to safety because the growing pack ice prevented their escape were never heard from again.

Long-term records of annual ice accumulations are decipherable from physical and chemical properties of the Greenland ice cores extracted from the depths of the ice sheet. The data reveal annual variations in the Earth's climate over hundreds of thousands of years. Information about modern patterns of precipitation and ice accumulation is derived from direct measurements of accumulation on the surface of the ice sheet and from snow pits and shallow ice cores.

Greenland's ice cores were found to have detectable traces of volcanic eruptions dating to 10,000 years ago, near the end of the last ice age. Direct temperature measurements were made by comparing the changing propor-

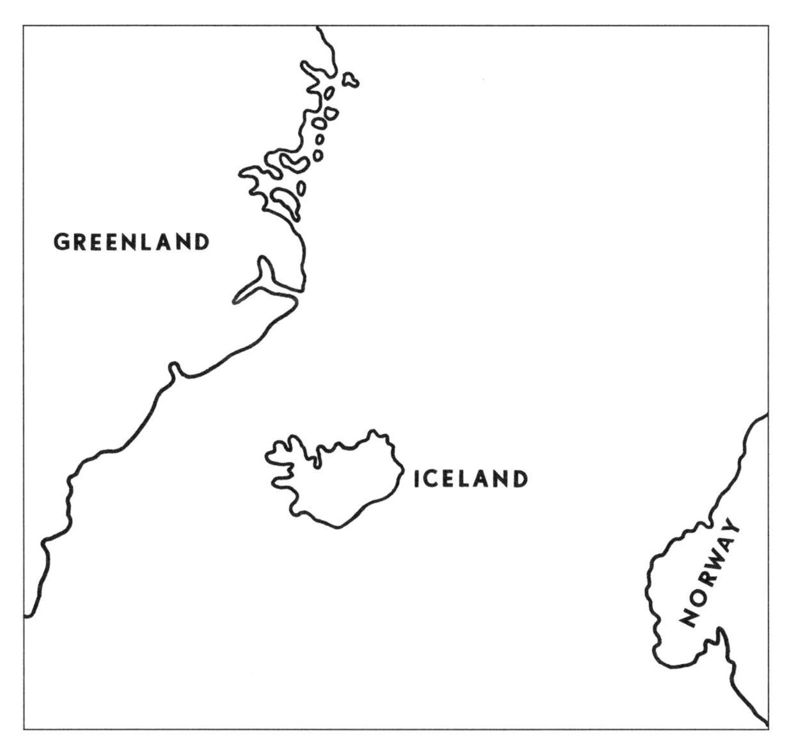

Figure 6–5 Location of Iceland in the North Atlantic.

tions of oxygen 16 (O_{16}) and oxygen 18 (O_{18}) isotopes in the ice. Water comprising the lighter of the two isotopes evaporates more easily. Therefore, ice containing a higher proportion of O_{16} compared to O_{18} means a colder climate because more heavy O_{18} remains behind in the ocean.

Volcanic eruptions eject acid gases into the atmosphere (Fig. 6–6) and periods of acid precipitation follow episodes of great volcanic activity. The traces of acid in the ice conduct electricity better than ordinary ice and are detectable by observing a rise in electrical conductivity through the ice core along its length. A drop in electrical conductivity indicates the presence of neutralizing bases, which are carried by windblown dust, reflecting dry, windy, and dusty conditions in the Northern Hemisphere at that time.

A comparison of the top layers of the ice with historical records of volcanic eruptions shows a good agreement. The acidity changes in the ice match other climatic indicators such as tree rings, whose narrow bands of growth indicate colder or drier seasons. They also correlate with written records of past climate.

The 1783 eruption of the Icelandic volcano Laki along with other major volcanic eruptions are recorded in ice cores taken from the Greenland ice sheet. The cores are accurately dated by counting the annual layers of ice. Those layers with a high acid concentration indicate times of major volcanic eruptions because volcanoes inject large amounts of sulfur compounds into the atmosphere, which precipitate as sulfuric acid. From the years 1500 B.C. to A.D. 1500, the cores clearly indicated five great European

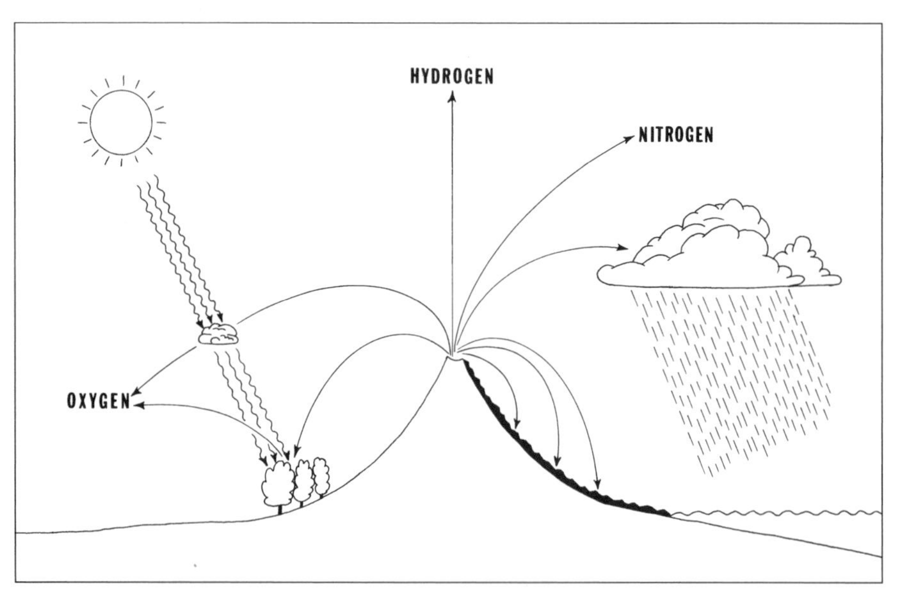

Figure 6–6 Volcanoes contribute large amounts of water vapor, carbon dioxide, and other important gases to the atmosphere as well as contribute to the growth of the continents.

eruptions, three in the Mediterranean area, and two in Iceland. The ice cores plainly show another massive volcanic eruption on Tambora, Indonesia, in 1815, which was the largest and produced the worst climatic effects of any volcano in modern history.

Climatologists argue that several volcanic eruptions occurring around the world in a short time span inject so much dust into the upper atmosphere that they alone could cause the onset of an ice age. Their argument is supported by geologic evidence of thin layers of volcanic dust buried in sediments correlating with times of increased ice cover. Volcanic debris has also been found in various layers of the Greenland and Antarctic ice cores, providing strong evidence that frequent and violent eruptions accompanied the ice age.

Eruptions on Iceland might have been more frequent during warm interglacial periods, when the island was largely free of ice. Over the last 300,000 years, four periods of intense volcanic activity, which spewed out up to hundreds of times more material than present volcanic eruptions, coincided with interglacial periods that punctuate the longer ice ages. The evidence is found in volcanic ash layers buried in sediments from the

Figure 6–7 Columbia Glacier, Prince William Sound, Valdez district, Alaska.
Courtesy of USGS

bottom of the Norwegian Sea. Apparently, the ice caps help quiet volcanism by bearing down on the Earth's crust, putting weight on the magma chambers that feed volcanoes. When the ice melts, the pressure on the chambers releases, causing eruptions, which suggest that volcanic activity is most active immediately following an ice age.

SURGE GLACIERS

Surge glaciers are like ordinary glaciers except for their diabolical behavior of occasionally dashing toward the sea at breakneck speeds. North America is host to about 200 surge glaciers, some of which seem destined to cross the Alaskan pipeline that brings oil from the North Slope to the seaport of Valdez. Ice cliffs of the Columbia Glacier (Fig. 6–7), located just west of Valdez, tower 300 feet above the deep water of the Prince William Sound.

For hundreds of years, the glacier has been calving icebergs and smaller chunks of ice called growlers, bergy bits, and brash ice that could make shipping difficult in the Sound. During the early 1980s, the glacier was in full retreat, moving rapidly backward more than 1.5 miles in 4 years. Many other glaciers have been retreating over the last 100 to 150 years (Fig. 6–8).

During most of their lifetimes, surge glaciers behave like normal glaciers, traveling at a snail's pace of perhaps a few hundred feet per year. However, at intervals of 10 to 100 years, the glaciers gallop forward up to 100 times

Figure 6–8 The Maclaren Glacier on the south side of the Alaskan Range is retreating, and moraines in front of the terminus indicate the extent of the glacier within the last few centuries. Photo by T. L. Pewe, courtesy of USGS

Figure 6–9 Avalanche on the Sherman Glacier, Cordova district, Alaska, caused by the August 24, 1964 Alaskan earthquake. Courtesy of USGS

faster than usual. A remarkable example is the Bruarjokull Glacier in Iceland, which in a single year advanced 5 miles, sometimes at an astonishing rate of 16 feet per hour.

A surge often progresses along the glacier like a great wave, proceeding from one section to another similar to the motion of a caterpillar. Subglacial streams of meltwater appear to act as a lubricant, allowing the glacier to flow rapidly toward the sea. Increasing water pressure under the glacier can lift it off its bed, overcoming the friction between ice and rock, thus freeing the glacier, which quickly slides downhill under the pull of gravity.

Surge glaciers also might be influenced by climate, volcanic heat, or earthquakes. Yet many of these glaciers exist in the same regions as ordinary glaciers, often almost side by side. Furthermore, the great 1964 Alaskan earthquake, which tore up much of the southern region, failed to cause more surges than before (Fig. 6–9).

The Bering Glacier in southern Alaska is the largest and longest surge glacier in North America, as well as the biggest temperate surging glacier on Earth. In June 1993, much of the glacier began speeding downslope for the first time in 26 years. The 120-mile-long glacier has experienced similar phases of rapid movement with up to 2 years duration around 1900, 1920, 1940, and the late 1950s, with the last one ending in 1967.

The surge began in the middle of the glacier and expanded until it reached the seaward end. Normally, the glacier moves at a rate of about 10 feet per day, but this last surge accelerated it up to 100 times greater than normal speed. In just three weeks, the glacier's terminus moved forward about a mile. As the glacier continued to surge, it displayed a variety of features, including deeply crevassed bulges and pressure ridges, exten-

Figure 6–10 Hubbard Glacier, Yakutat district, Alaska. Photo by A. Post, courtesy of USGS

sional fractures with intersecting crevasses, tear faults, grabens, and other stress-generated structures. In July 1994, an outburst flood began from under the face of the margin of the glacier, and a large discharge of sediment-laden water and huge blocks of ice spouted forth.

The Hubbard Glacier (Fig. 6–10) reaches out of the rugged St. Elias Mountains toward the Gulf of Alaska near the Canadian border. About 800 years ago, Hubbard stampeded toward the sea, retreated, and advanced again 500 years later. Since 1895, the 70-mile-long river of ice has been steadily flowing toward the Gulf of Alaska up to 200 or more feet per year. Then in June 1986, the glacier galloped ahead as much as 47 feet a day.

Meanwhile, Valerie Glacier, a western tributary, advanced up to 112 feet per day. Extraordinarily heavy snows in southeastern Alaska during the three previous winters might have forced Hubbard's advance. The surge closed off Russell Fjord with a formidable ice dam, some 2,500 feet long and up to 800 feet high, whose impounded waters threatened the town of Yakutat to the south.

As many as 20 similar glaciers around the Gulf of Alaska are charging toward the sea. Once they reach the ocean, they would raise sea levels. West Antarctic ice shelves, which are floating portions of glaciers that have flowed into the sea and are loosely anchored by submerged islands, could rise off the seabed and drift away, causing a flood of ice to surge into the sea around Antarctica.

Much of the West Antarctic ice sheet rides on glacial till. Water pressure in the till below a stationary ice stream is lower than it is below a moving ice stream. Therefore, the higher the pressure, the more easily the ice floats and the faster it moves. With a continued rise in sea level, more ice would plunge into the ocean, raising sea levels even higher, which would release more ice and set in motion a vicious circle. The additional sea ice floating toward the tropics would also increase the Earth's albedo and lower global temperatures.

The ice would chill the waters in the Antarctic, and ocean currents would distribute the cold water northward. One such current originating in the Antarctic sends cold, salty bottom water flowing as far north as New Jersey. The drop in ocean temperatures could dramatically change the path of ocean currents and disrupt weather patterns significantly to start a new ice age. This scenario appears to have been staged at the end of the previous warm interglacial, when sea ice dramatically cooled the ocean, initiating the last ice age.

ICEBERGS

At the peak of the last ice age, excess ice at both poles calved off into the ocean to form icebergs. Massive glaciers flowing into the sea produced intense iceberg activity, and icebergs covered up to half the area of the oceans. Like ice cubes in a cold drink, they cooled the ocean surface and weakened oceanic circulation. Near the end of the ice age, so many icebergs clogged the North Atlantic, they caused the retreating glaciers to pause in mid stride, and global temperatures temporarily returned to ice age conditions.

During the warmer intervals of the last ice age, large armadas of icebergs broke off the continental glacier over Canada and invaded the North Atlantic from Labrador to Europe. Every time a mountain of ice up to 2 miles high formed over Hudson Bay, it became so thick it blocked heat escaping from the Earth's interior. As a result, the bottom layer of the ice melted, leading to a catastrophic movement of the ice sheet, which calved off into great fleets of icebergs. The melting of the iceberg armadas flooded the ocean with fresh water, disrupting the warm Gulf Stream currents and thus chilling the northern climate.

About a trillion tons of ice discharges into the seas surrounding Antarctica annually, and glaciers flowing into the ocean calve off forming icebergs. The ice floats because unlike most other substances it is lighter as a solid than as a liquid because water expands upon freezing. Antarctic ice shelves are surrounded by floating ice pinned in by small, isolated islands buried beneath the ice. This creates an unstable ice mass that could account for the large number and size of icebergs breaking off the Ross Ice Shelf.

The icebergs appear to be getting larger, possibly due to a warmer global climate. If the apparent global warming continues, Antarctic ice sheets could destabilize and break off into the ocean, creating additional sea ice. The increased area of sea ice could form a gigantic ice shelf, covering up to 10 million square miles, or nearly three times greater than the United States.

The number of extremely large icebergs has also increased dramatically. The Antarctic consists of several ice shelves, which are thick sheets of floating ice that have slid off the continent. Slowly, the continental ice sheet flows down to the sea, where portions break off to form icebergs. The largest known iceberg separated from the Ross Ice Shelf in late 1987 and measured about 100 miles long, 25 miles wide, and 750 feet thick, about twice the size of Rhode Island. In August 1989, it collided with Antarctica and broke in two. Another extremely large iceberg broke off the Antarctic ice sheet in early March 1995 and headed into the Pacific Ocean.

An abundant source of fresh water that has been largely overlooked is locked up in icebergs. Every year, some 5,000 icebergs, totaling about 200 cubic miles of ice, calve off Antarctica. The icebergs generally have flat tops and steep sides (Fig. 6–11) and consist of approximately 100 million tons of fresh water. The icebergs drift along with the wind and currents for many years, finally breaking up and melting when entering warmer water. Occasionally, a large iceberg drifts as far north as 30 degrees south latitude, where it might influence the regional climate as well as the local weather.

Figure 6–11 A large flattop iceberg off Antarctica. Courtesy of U.S. Navy

Studies have shown the feasibility of capturing icebergs and towing them to Australia, South Africa, Saudi Arabia, and California, utilizing them as a supply of fresh water. Only very large icebergs could be used in this manner because smaller icebergs tend to break up even in moderate seas. Furthermore, because of melting and erosion, a larger iceberg improves the chance of a substantial block of ice making it to its destination.

When arriving at a seaport, the iceberg would be mined using existing technology. A slurry of ice and water would be pumped ashore through an ocean pipeline. However, such a scheme is not without its pitfalls, not only because of accidents resulting from broken off icebergs in the shipping lanes, but also due to the uncertainty of the environmental consequences of moving large numbers of icebergs around the world.

THE RISING SEA

Glacial ice covers nearly one-tenth of the Earth's surface, and glaciers contain about 70 percent of all the fresh water in the world. Alpine or valley glaciers exist on every continent and hold as much fresh water as all the world's rivers and lakes. If the ice caps melted during a sustained warmer climate, they could substantially raise sea levels and drown coastal regions. Beaches and barrier islands would disappear as shorelines move inland (Fig. 6–12).

If the present rate of melting continues, the sea could rise a foot or more by the year 2030, comparable to the rate of melting at the end of the last ice age. If all the ice melted, the additional seawater would move the shoreline up to 70 miles inland in most places and much more at low-lying areas such as river deltas. The inundation would in turn radically alter the shapes of the continents. Mississippi, Louisiana, east Texas, and major parts of Alabama and Arkansas would practically disappear. All of Florida along with south Georgia and the eastern Carolinas also would be gone. Most of the Isthmus separating North and South America at Panama would sink out of sight.

For every foot of sea level rise, up to 1,000 feet of seashore would disappear, depending on the slope of the coastline. Just a 3-foot rise could flood about 7,000 square miles of coastal land in the United States, including most of the Mississippi Delta, possibly reaching the outskirts of New Orleans. Other parts of the world would fare much worse. Half the scattered islands of the Republic of Maldives, southwest of India, would be lost. Much of Bangladesh also would drown, a particularly distressing situation since the people there can barely support themselves off the existing land.

Figure 6–12 Serious losses of property near Cape Hatteras, Dare County, North Carolina, caused by shoreline regression and storm surges. Photo by R. Dolan, courtesy of USGS

TABLE 6–1 MAJOR CHANGES IN SEA LEVEL

Date	Sea Level	Historical Event
2200 B.C.	low	
1600 B.C.	high	Coastal forest in Britain inundated by the sea.
1400 B.C.	low	
1200 B.C.	high	Egyptian ruler Ramses II builds first Suez Canal.
500 B.C.	low	Many Greek and Phoenician ports built around this time are now under water.
200 B.C.	normal	
A.D. 100	high	Port constructed well inland of present-day Haifa, Israel.
A.D. 200	normal	
A.D. 400	high	
A.D. 600	low	Port of Ravenna, Italy, becomes landlocked. Venice is built and is presently being inundated by the Adriatic Sea.
A.D. 800	high	
A.D. 1200	low	Europeans exploit low-lying salt marshes.
A.D. 1400	high	Extensive flooding in Low Countries along the North Sea. The Dutch begin building dikes.

Down through the centuries, civilizations have had to cope with changing sea levels (Table 6–1). The Dutch, who worked so hard to reclaim their land from the sea, would find a major portion of their country lying under water. Many islands would drown or become mere skeletons of their former selves with only their mountainous backbones showing above the water. Most of the major cities of the world, because they are either located on seacoasts or along inland waterways, would be inundated by the sea with only the tallest skyscrapers poking above the waterline.

The receding shores would result in the loss of large tracks of coastal land along with shallow barrier islands. Delicate wetlands, where many species of marine life hatch their young, would simply vanish (Fig. 6–13). Low-lying fertile deltas that feed a large portion of the world's population would

Figure 6–13 The Mosquite Lagoon, Cape Canaveral National Seashore, Florida.
Courtesy of National Park Service

be inundated by the rising waters. Coastal cities would have to move farther inland or build costly seawalls to keep out the ocean.

Crashing waves accompanying storms at sea cause serious erosion of sand dunes and sea cliffs. The constant pounding of the surf also erodes most man-made defenses against the rising sea. Upward of 90 percent of America's once sandy beaches are sinking beneath the waves. Barrier islands and sandbars running along the east coast and east Texas are disappearing at alarming rates. Beaches along North Carolina are retreating at a rate of 4 to 5 feet per year. Sea cliffs are eroding back several feet a year, often destroying expensive homes. Most defenses used in a fruitless attempt to stop beach erosion usually end in defeat as waves relentlessly batter the coast.

The global sea level appears to have risen upward of 6 inches over the last 100 years (Fig. 6–14) due mainly to the melting of the Antarctic and Greenland ice sheets. In some areas like Louisiana, the level of the sea has risen as much as 3 feet per century. The thermal expansion of the ocean has also raised the level of the ocean about 2 inches. Surface waters off the California coast have warmed nearly 1 degree over the past half century, causing the water to expand and raise the sea level about 1.5 inches. Higher sea levels are also due in part to the sinking of the land because of the

Figure 6–14 Sea level through time.

increased weight of seawater pressing down on the continental shelf. Sea level measurements are further affected by the rising and sinking of the land surface due to plate tectonics, and, at the end of the last ice age, were altered by the rebounding of the land after the melting of the great ice sheets.

The rapid deglaciation at the end of the last ice age between 14,000 and 7,000 years ago, when torrents of meltwater entered the ocean, raised the sea level on a yearly basis just a few times greater than it is rising today. The present rate of sea level rise is several times faster than it was 40 years ago, amounting to about an inch every 8 years. Most of the increase appears to be the result of melting ice caps. The extent of polar sea ice appears to have shrunk as much as 6 percent during the decades of the 1970s and 1980s. Alpine glaciers, which contain large quantities of ice, appear to be melting as well (Fig. 6–15).

One method to determine whether the ocean is actually warming is by measuring the speed sound waves travel through the sea. Since sound travels faster in warm water than in cold water, a phenomenon known as acoustic thermometry, a decade of measurements could reveal whether global warming is a certainty. The idea is to transmit low-frequency sound waves from two stations—one 26 miles off the coast of Point Sur, California, and the other 8 miles from Kauai, Hawaii,—and monitor them from 12 listening posts scattered around the Pacific from Alaska to New Zealand. The sound sources would lie some 2,800 feet beneath the sea to minimize their effects on marine life. The signals would take several hours to reach the farthermost stations. Therefore, shaving a few seconds off the travel time over an extended period of a decade or more could indicate that the oceans are indeed warming.

The first possible signs that rising global temperatures have started to warm the ocean were revealed in satellite measurements of the extent of the polar sea ice. Sea ice forms a frozen band around Antarctica and covers most of the Arctic Ocean during the winter season in each hemisphere. The total surface area of the ice does not appear to have changed significantly, but the maximum extent the ice pack reaches outward from the poles during the winter has diminished.

Figure 6–15 The Gulkana Glacier, Donelly district, Yukon region, Alaska, has been shrinking in recent decades. Photo by M. F. Meier, courtesy of USGS

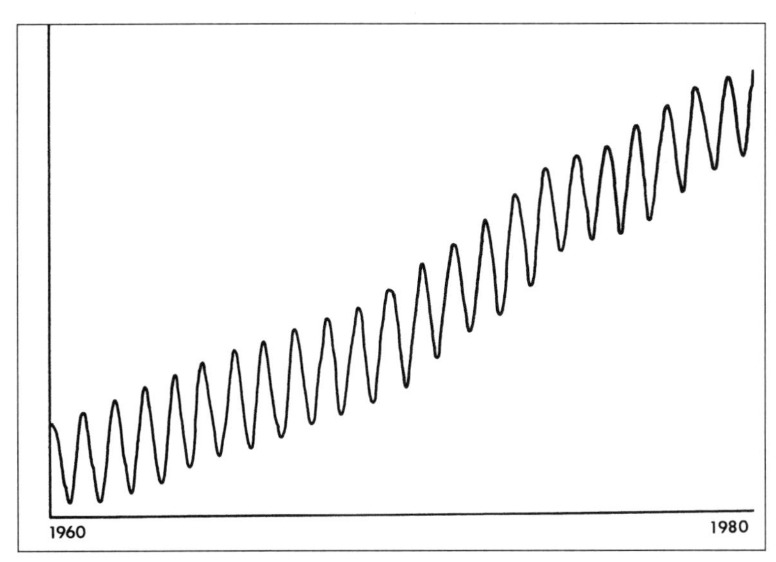

Figure 6–16 Seasonal and long-term variations in atmospheric carbon dioxide levels.

The ice obtains its maximum extent during the spring in the Southern Hemisphere, when the Antarctic ice is breaking up and the Arctic ice is beginning to spread. Sea ice hinders heat flow from the warm ocean to the cold atmosphere, affecting the global climate system. As the ocean continues to warm, the ice melts closer to the poles, further reducing the perimeter of sea ice. However, an extensive study of temperatures over the Arctic Ocean indicates the region has not warmed significantly over the last 4 decades, perhaps because the Arctic is the last region affected by global warming.

If average global temperatures continued rising due to an increase in atmospheric carbon dioxide levels (Fig. 6–16), the present interglacial could become equally as warm as the last one. The warmer climate could induce an instability in the West Antarctic ice sheet, causing it to surge into the sea. This rapid flow of ice into the ocean could raise sea levels and inundate the continents.

During the Eemian interglacial prior to the last ice age, the warmer seas melted the polar ice caps, raising the level of the ocean about 60 feet higher than today. Yet the warming failed to prevent the onset of the last glaciation, suggesting that today's warming trend could trigger the onset of another ice age. Indeed, according to the theory of orbital variations along with geologic evidence secured from rock cores of the seabed and from ancient coral growth patterns, the next ice age is long overdue.

7

THE ARCTIC

The Arctic is the region surrounding the North Pole above about 67 degrees north latitude. It includes all the extreme northern lands surrounding the Arctic Ocean: the upper regions of Alaska (Fig. 7–1), Canada, most of Greenland, the northern tip of Iceland, and the northlands of Scandinavia and Russia. During the summer months, the region is bathed in sunlight 24 hours a day, and in the winter months it is cloaked in 24-hour-a-day darkness.

The American Admiral Robert Peary has long been credited with first reaching the North Pole, on April 6, 1909 (Fig. 7–2), although some controversy remains as to whether he actually arrived at 90 degrees north latitude, due to possible navigational errors. In August 1958 the American nuclear submarine USS *Nautilus* was the first vessel to travel underneath the Arctic ice pack from the Bering Sea to the Greenland Sea. On August 22, 1994, the U.S. Coast Guard ship *Polar Sea* and the Canadian Coast Guard ship *Louis St. Laurent* became the first surface vessels to reach the North Pole.

Figure 7–1 The northern foothills and Arctic coastal plain province east of the Kukpowruk River, Utukok-Corwin region, northern Alaska. Photo by R. M. Chapman, courtesy of USGS

ARCTIC TUNDRA

Only about a third of the Earth's land surface, or roughly 20 million square miles, is wilderness with little signs of human presence, including roads, settlements, buildings, airports, railroads, pipelines, power lines, dams, reservoirs, and oil wells. Except for a few scattered outposts, the ice continent of Antarctica is practically all wilderness, a situation that could well change as nations begin to explore for petroleum and minerals.

Several broad belts of wilderness wind around the globe. One band stretches across the Arctic tundra of northern Alaska (Fig. 7–3a and b), Canada, and the northernmost reaches of Eurasia. Another runs southwest from far eastern Asia through Tibet, Afghanistan, and Saudi Arabia into

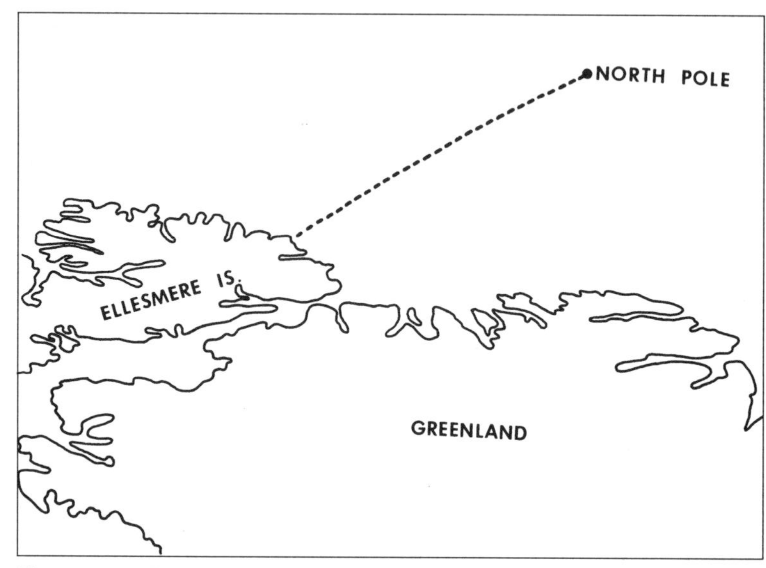

Figure 7–2 Peary's route to the North Pole.

Africa. The forbidding Sahara Desert in northern Africa and the great central desert of Australia serve as additional wilderness areas.

Wild patches also exist in other parts of Africa, around the Amazon, and along the Andes Mountains of South America. Less than 20 percent of the

Figure 7–3a Arctic tundra in southwestern Copper River Basin, Alaska. Photo by J. R. Williams, courtesy of USGS

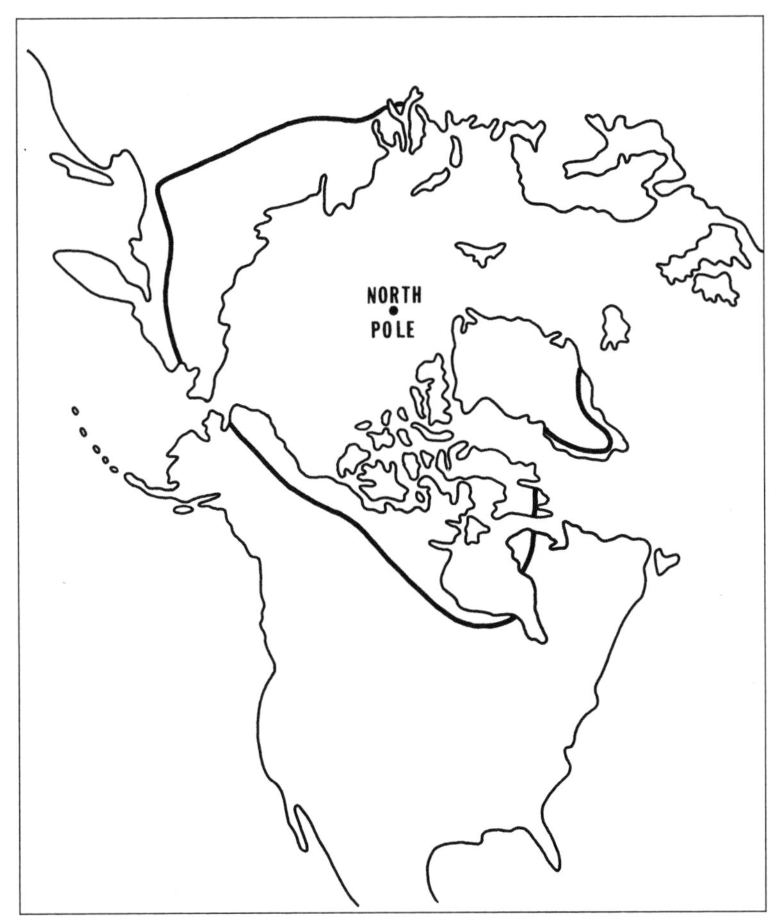

Figure 7–3b The Arctic tundra line, north of which the ground is frozen year-round.

identified wilderness areas is legally protected from exploitation. At least half the remaining wildlands is not self-protecting by virtue of their forbidding natures and could easily be destroyed as billions of more people are added to the world's population.

One of the most barren environments is the Arctic tundra of North America and Eurasia. It covers about 14 percent of the world's land surface in an irregular band winding around the top of the world, north of the tree line and south of the permanent ice sheets. Alpine tundra (Fig. 7–4) covers much of the world's mountainous terrain above the tree line and below mountain glaciers.

Unlike the Arctic tundra, which lies at high latitudes and therefore is deprived of sunlight during the long winter months, alpine tundra receives daily doses of sunlight. While little snow falls in much of the Arctic, alpine areas receive abundant snowfall because of their high elevations. The vegetation in both regions has much in common, however, consisting mostly of stunted plants, often widely separated by bare soil or rock.

The Arctic tundra is also one of the most fragile environments in the world, and human activity can cause much damage. The polar front is the boundary between polar and tropical air masses and is associated with the jet stream. The jet stream moves with the polar front farther to the south in winter and farther to the north in summer. During the winter, when the polar front sweeps across polluted regions of the Northern Hemisphere, it removes atmospheric pollution and transports it to Arctic regions, contaminating the once pristine skies and producing a phenomenon known as Arctic haze. The haze makes the Arctic as polluted in winter and early spring as anywhere in North America and Eurasia.

In the past 40 years, the haze has become a near permanent feature of the far north. The shade resulting from the haze might explain why the Arctic has warmed much less than the Antarctic in the last few decades. However, the haze also traps what little heat the region receives from the sun, causing the Arctic to actually warm. The smog, which originates mostly in Europe and northwest Asia and is often as bad as air pollution in some American suburbs, blocks out sunlight, which is at a premium in these high latitudes.

Figure 7–4 Alpine tundra on Gunsight Mountain, Alaska. Photo by J. R. Williams, courtesy of USGS

Figure 7–5 A precipitation and air temperature gauge at Wolverine Glacier, Alaska. Photo by W. V. Tangborn, courtesy of USGS

The lack of significant precipitation to scrub out pollutants from the atmosphere allows the smog to remain for extended periods.

The acidity level of the precipitation can be quite harmful to delicate tundra flora. Lichens live in areas where no other plants exist and grow very slowly. Because they absorb dangerous substances from atmospheric moisture and dust that can build to lethal levels, they are usually among the first to perish. Animals feeding on the contaminated plants accumulate the poisons in their bodies, which could prove to be an ecological disaster.

Much of the tundra is barren mainly due to limited food resources, high winds, and frigid temperatures for most of the year (Fig. 7–5). Parts of the Arctic tundra are the world's most nutrient-poor habitats, and yet algae and aquatic insects proliferate in the cold streams that flow into the Arctic Ocean. In this land of unusual climatic conditions, plants and animals must make the best of limited growing season, rainfall, and nutrients. The growing season in the tundra is generally only 2 to 3 months long. But a slight increase in temperature due to global warming would increase the growing season, causing changes in biological communities.

Special survival adaptations are required in this demanding environment, where sturdy as well as frail species live. A classic example are seeds of the Arctic lupine, collected from an ancient lemming burrow that dated near the end of the last ice age, about 10,000 years ago. Perhaps the most remarkable quality about these seeds is how well they were preserved, and when planted they actually grew, blossoming with delicate flowers. Although other ancient seeds such as wheat and corn found in Egyptian tombs and elsewhere have been grown successfully, these are by far the oldest seeds to have ever germinated, an indication of how hardy tundra species are.

Despite its seeming hardiness, the Arctic tundra is at risk (Fig. 7–6). Overgrazing by reindeer on the sparse grassland can decimate large areas. Exploration activities for oil and minerals can ruin huge acreages, and tracks of vehicles traveling cross-country are still visible a half century later.

The boreal forest is a vast band of conifers and other softwoods stretching across northern Eurasia and North America (Fig. 7–7). Over the past century, the forest has absorbed large quantities of excess carbon dioxide from industrial activities. Unfortunately, due to logging and massive increases in tree dieback from fires, acid rain, and diseases, mostly due to warmer weather in much of the Arctic since the 1970s, the absorption of carbon dioxide has been limited.

Figure 7–6 Gravel road near Umiat showing severe subsidence caused by thawing of ice-wedge polygons in permafrost, Anaktuvik district, Northern Alaska. Photo by O. J. Ferrians, courtesy of USGS

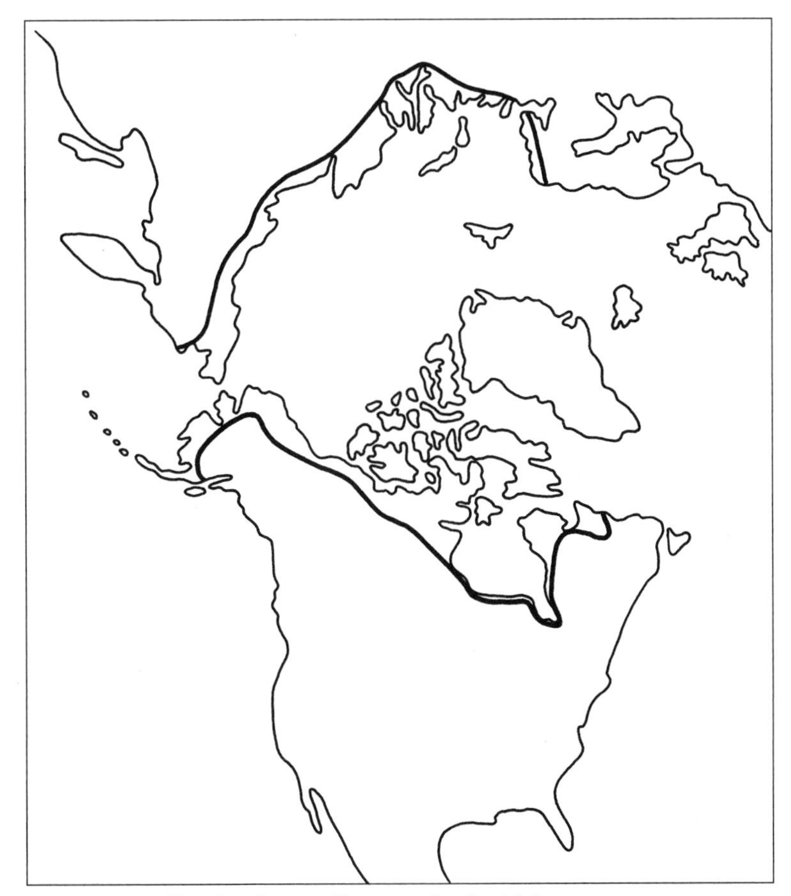

Figure 7–7 The Arctic tree line, north of which no trees grow.

Arctic species tend to be poor competitors, and as trees migrate northward in response to an apparent global warming, Arctic ecosystems could decline. The effects of global warming could last for centuries, during which forests would creep northward and other wildlife habitats including the Arctic tundra would disappear entirely. The northward-shifting habitats might cause some tundra species to become extinct due to the breakup of mixed habitats into more uniform ecosystems.

The rate of change might be too great for many ecosystems to adjust rapidly enough. The flat, open tundra, where only lichens and other low-growing plants live, would witness an incursion of taller-growing plants, which would in turn affect the nesting behavior of Arctic-breeding shorebirds. Fish and marine mammals living in the Arctic waters also could be adversely affected by a change in the global climate.

A major concern over increasing amounts of atmospheric pollution is global warming from the trapping of escaping thermal energy by man-made greenhouse gases. Higher global temperatures, which are amplified in the high latitudes, might thaw the Arctic tundra. The warming would release into the atmosphere vast quantities of carbon dioxide and methane trapped in the soil by decomposing plant material.

Increased carbon dioxide also fertilizes wetlands, causing them to emit more methane. These processes could produce a runaway greenhouse effect, raising global temperatures to lethal levels. Increased cloudiness and precipitation in the high-latitude regions of the Northern Hemisphere would occur, and permafrost would begin to slowly disappear.

Evidence collected from Alaska's northern tundra suggests that global warming might already be releasing carbon dioxide from the land. Boreholes drilled across the Arctic tundra of northern Alaska show anomalous warming in the upper 300 to 500 feet of permafrost and rock. The warming is between 2 and 4 degrees Celsius over this century, much greater than the global average warming of perhaps 1 degree over the same period because the effects of global warming are greatly magnified in the Arctic regions.

Figure 7–8 An ice wedge in permafrost exposed by placer mining near Livengoos, Tolovana district, Yukon region, Alaska. Photo by O. J. Ferrians, courtesy of USGS

The Arctic climate varies more than it does elsewhere, making the detection of changing trends difficult. However, temperature data from Arctic landmasses do show a warming trend over the last three decades. The Alaskan tundra has also become a net source of carbon dioxide, releasing more of this gas during the summer than it absorbs.

PERMAFROST

Most of the ground in the Arctic tundra is permafrost (Fig. 7–8), meaning it is frozen year-round, and only the top few inches of soil thaws during the short summer season. Even though sunlight bathes the ground 24 hours a day, the soil temperature seldom rises much above the freezing point because most of the radiant energy is spent melting the ice in the soil. Often over a lengthy period, this freeze-thaw sequence produces bizarre patterns

Figure 7–9 A type of solifluction called earthruns on a hill in Nome district, Seward Peninsula, Alaska. Photo by P. S. Smith, courtesy of USGS

Figure 7–10 Deformation of a wood bridge by frost action, causing the cross brace to split by upward thrusting of the pile. Photo by T. L. Pewe, courtesy of USGS

in the ground. Some features consist of regular polygonal shapes produced when the ground heaves up due to the expansion of ice as it freezes.

In the tundra environment, severe subsidence can occur in permafrost regions. Solifluction is a type of mudflow occurring in colder climates. It is a slow downslope movement of water-logged sediments that causes ground failures in permafrost (Fig. 7–9). When frozen ground melts from the top down during warm spring days in the temperate regions or during the summer in areas of permafrost, it causes surface mud to move downslope over a frozen base on whose surface it easily slides. This ground motion creates many problems in construction, especially in the far northern permafrost areas. Foundations must extend down to the permanently frozen layers, or whole buildings could be damaged by the loss of support or by lateral movement downslope.

Soil creep is the slow downslope movement of overburden and bedrock. It might be very rapid where frost action is prominent. After a freeze-thaw sequence, material moves downslope due to the expansion and contraction of the ground. Another type of movement of soil material associated with cycles of freezing and thawing is called frost heaving. It thrusts boulders and other structures upward through the soil and causes serious construction problems (Fig. 7–10).

Frost action can produce mechanical weathering by exerting pressures against the sides of cracks and crevices in rocks when water freezes inside them, resulting in frost wedging (Fig. 7–11). This widens the cracks, while surface weathering tends to round off the edges and corners, providing a landscape resembling multitudes of miniature canyons up to several feet wide carved into solid bedrock.

At the margins of glaciers across the northern tundra are rugged periglacial regions. The term refers to the conditions, processes, and topographic features of areas adjacent to the borders of a glacier. A periglacial climate is characterized by low temperatures that fluctuate about the freezing point and strong wind action during certain times. Periglacial processes directly controlled by the glacier sculpted features along the tip of the ice. Cold winds blowing off the glaciers affected the climate of the glacial margins and helped create periglacial conditions. The zone is dominated by such processes as frost heaving, frost splitting, and sorting, which created immense boulder fields out of once solid bedrock.

The Arctic tundra offers many challenges to construction. The discovery of oil at Alaska's North Slope resulted from intensive exploration for new reserves. The Russian tundra, about 1,000 miles northeast of Moscow, has witnessed haphazard oil-drilling activity with little regard to the environment, resulting in massive oil spills that are difficult to clean up due to the extreme cold.

Figure 7–11 A frost-split block of granite on moraine near Moraine Lake, Sequoia National Park, Tulare County, California. Photo by F. E. Matthes, courtesy of USGS

Figure 7–12 DEW line radomes and transmitters at Tatalina Air Force Station in Alaska. Courtesy of U.S. Air Force

By far the largest project across the Arctic tundra was the construction of the Distant Early Warning (DEW) line in the 1950s. The system for detecting Russian bombers during the Cold War comprised a chain of 33 radar stations some 250 miles north of the Arctic Circle. The DEW line stretched 3,400 miles from Point Barrow, Alaska, across the northern reaches of Canada's uncharted frozen tundra, finally ending on the east coast of Greenland.

Each station had to endure the rigors of the Arctic winter, when subzero weather and 100 mile-per-hour winds played havoc with both crews and equipment (Fig. 7–12). In addition, giant ocean radar platforms similar to offshore drilling rigs, called "Texas towers," were constructed to fill in the gaps between the Arctic seas and the landmasses. They later had to be dismantled because violent Arctic storms were ripping them apart.

The Arctic tundra preserves ancient meteorite craters better than most other environments. One example is the New Quebec Crater in Canada, the largest known meteorite crater where actual meteoritic debris has been found. It measures about 11,000 feet in diameter and about 1,300 feet in depth. The crater contains a deep lake, the surface of which is 500 feet below the crater rim. The impact structure is relatively young, estimated at only a few thousand years old. It is one of the most well-preserved craters because few changes occur in the frigid tundra.

THE POLAR VORTEX

The primary function of the general circulation of the atmosphere is to distribute heat from the tropics to the polar regions (Fig. 7–13). Air currents travel from the equator to the poles and back again in cellular structures called Hadley cells, named for the 18th-century meteorologist George Hadley, who first recognized atmospheric convection.

The rotation of the Earth prevents the formation of a single large equator-to-pole cell. Instead, the circulation is achieved by three separate cells in each hemisphere, transferring heat and cold from one cell to the next. The interaction between cells is complicated by the distribution of oceans, continents, mountain ranges, deserts, forests, and glaciers. The polar vortexes are another form of cell structure. They are easterly blowing polar winds that circle the Earth around the North and South poles.

Every September and October since the late 1970s, a giant hole opens up in the ozone layer over Antarctica, where half the ozone disappears. A similar ozone hole exists at times over the Arctic as well, where the loss of winter ozone is as much as 15 percent. The polar vortex traps cold air over each pole, where water vapor condenses into clouds of ice (Fig. 7–14) that accelerate ozone destruction by chemical reactions. The ozone hole also cools the air inside the polar vortex, producing additional ice clouds in a

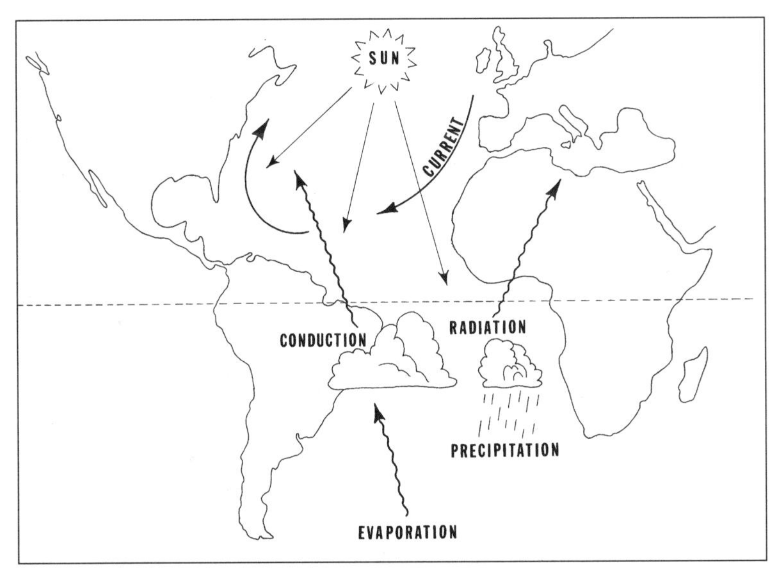

Figure 7–13 Heat flow between the ocean and atmosphere.

Figure 7–14 Polar stratospheric clouds north of Stavanger, Norway. Courtesy of NASA.

feedback mechanism. When the ozone holes break up in the spring, the polar regions export their depleted ozone air along with ozone-destroying chemicals to the midlatitudes, which experience increased ultraviolet radiation levels.

The ozone depletion is strongly believed to be caused by man-made chemicals as well as natural causes such as volcanic eruptions. Chlorine monoxide in air pollution destroys ozone molecules in the Antarctic stratosphere, where the chemical can be 100 times normal levels. Polar stratospheric clouds composed of frozen water and nitric acid crystals help chlorine destroy the ozone layer by promoting chemical reactions that compete for the free oxygen atoms that normally combine with oxygen molecules to produce ozone. For this reason, the Arctic ozone hole is less well defined than that in the Antarctic because temperatures generally are not cold enough to allow the formation of cloud crystals.

The sunspot cycle is a periodic recurrence of sunspots, ranging in duration from 9 to 14 years, with an average of about 11 years. Other solar activity varies directly with the solar cycle, which has a period of about 22 years. In addition, the polarity of the sun's magnetic field reverses every other sunspot cycle, or about every 22 years. The sunspots associate with

Figure 7–15 A spectacular aurora in the Southern Hemisphere from Spacelab aboard the space shuttle. Courtesy of NASA

solar magnetic fields that are several thousand times stronger than the magnetic field on the Earth's surface.

Perhaps the strongest link between solar activity and the climate is the effect sunspots have on the vortex of stratospheric winds that swirl over the North Pole during the winter. The polar vortex breaks down when the wind in the lower stratosphere over the equator changes direction from west to east. However, it only does this during a maximum number of sunspots. The more sunspots that mar the sun's surface, the warmer the wintertime temperatures due to the breakdown of the polar vortex and the subsequential intrusion of warm air.

The charged particles in the solar wind are funneled into the polar regions of the Earth's magnetic field and produce the magnificently colored lights of the aurora borealis in the Northern Hemisphere and the aurora australis in the Southern Hemisphere (Fig. 7–15), also called northern and southern lights. They are caused by the emission of light from atoms in the uppermost atmosphere excited by cosmic rays. A strong solar wind also might be responsible for the formation of noctilucent clouds composed of minute ice crystals about 50 miles altitude, far above normal clouds, making them the highest clouds on Earth.

Every summer since 1981, research radars have detected unexplained high-frequency echoes above the North and South poles. Over the same period, noctilucent clouds, which streak the high polar atmosphere called the mesosphere just after dark on summer nights, have been growing steadily brighter and becoming more common than usual. Apparently, the mesosphere is starting to feel the far-reaching effects of global warming. Atmospheric methane, which breaks down by sunlight into water molecules in the mesosphere, doubled since 1900, while noctilucent clouds grew nearly 10 times brighter.

THE ARCTIC OCEAN

Most of the seawater surrounding the continents lies in a single great basin in the Southern Hemisphere, which is nine-tenths ocean. It branches northward into the Atlantic, Pacific, and Indian basins in the Northern Hemisphere, which contains most of the landmass. The Atlantic and Pacific are in contact with both the Antarctic and the Arctic oceans. The Arctic Ocean is a nearly landlocked sea connected to the Atlantic and Pacific only by narrow straits.

About 20 million years ago, a ridge near Iceland subsided, allowing cold water from the recently formed Arctic Ocean to surge into the Atlantic Ocean, giving rise to the present oceanic circulation system. The drifting of the continents radically changed patterns of ocean currents, whose access to the poles was severely restricted, allowing the growth of glacial ice. When the continents drifted toward the poles, they blocked poleward oceanic heat from the tropics and replaced heat retaining water with easily chilled land. As the cooling progressed, the northern continents accumulated snow and ice, which reflected more sunlight out to space, causing further heat loss.

The Arctic Ocean is the least studied of the world's oceans and the only one that is practically landlocked. About 4 million years ago, Alaska approached Asia at the Bering Strait, practically closing off the Amerasian Basin from warm water currents originating from the tropics, resulting in the formation of a permanent ice cap over the North Pole. The narrow, shallow Bering Strait that separates Alaska from Asia also blocked the flow of deep, cold water from the Arctic Ocean into the Pacific. The closing off of the Arctic Ocean from warm Pacific currents might have provided the geographic conditions necessary for the development of the Pleistocene ice ages, when massive glaciers swept out of the polar regions and buried the northern lands. During the last ice age, the Arctic Ocean apparently was covered by an ice shelf hundreds of feet thick that killed off all life in the icy waters below.

The Arctic Ocean is a sea of pack ice with occasional ice islands large enough to support research stations. The medium extent of the Arctic ice cap is approximately 65 degrees north latitude. Sea ice covers an average area of about 4 million square miles, with an average thickness of 15 to 20 feet. Pack ice in the Arctic Ocean varies from over 3 million square miles in summer to nearly twice that amount in winter.

The world's ocean is constantly in motion, distributing water and heat to all corners of the globe. In effect, it is a huge circulating machine that makes the Earth's climate equitable. The cold, heavy Arctic and Antarctic waters flow near the bottom toward the equator. Along the way, the ocean currents follow well-defined courses, transporting tremendous quantities of seawater, which serves as a global "conveyor belt" over the planet (Fig. 7–16).

Eddies of swirling warm and cold water accompany the ocean currents. Many eddies are enormous, as much as 100 miles or more across and reach depths of 3 miles. Most are less than 50 miles across, and some, including those in the Arctic Ocean off Alaska, are only 10 miles wide. Like giant eggbeaters, eddies play an important role in mixing the oceans.

In the polar regions, the air near the ground is generally too cold to melt ice crystals falling from clouds, which precipitate as sleet or snow, building up glacial ice. The existence of ice at both poles established a unique equator-to-pole oceanic and atmospheric circulation system. In the absence of warm ocean currents flowing from the tropics to keep the polar regions free of ice, glaciers with their high albedo continue to exist.

Seawater freezes more readily in the Arctic regions because the near-surface water is not sufficiently enriched in salt and therefore not dense enough to sink. If the Bering Strait became totally blocked with ice, it would effectively prevent the warm Pacific current from entering the Arctic Ocean, which could keep it frozen and unnavigable year-round.

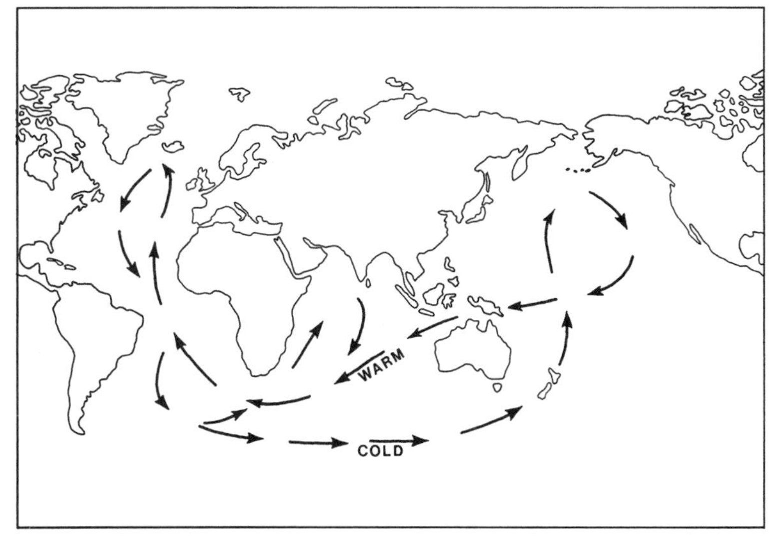

Figure 7–16 The ocean's heat transport system distributes warm water from the tropics to the cold regions of the world.

Figure 7–17 Gulf of St. Lawrence, Canada, from the *Skylab* space station, showing much ice and snow. Courtesy of NASA

Land surrounding the Arctic Ocean prevents warm ocean currents from reaching the North Pole and melting the polar ice cap. Even if the polar ice cap melted, the Arctic Ocean would not stay ice free for very long. With the present pole positions and atmospheric circulation, an iceless Arctic Ocean would induce heavy precipitation that would initiate a glacial advance. While the glaciers are on the move, the Arctic Ocean would freeze, cutting off much of the supply of moisture. Precipitation would continue until the North Atlantic became too cold to provide sufficient moisture. Melting would commence and continue until the Arctic Ocean was again ice free, setting the stage for the cycle to begin anew.

Before the age of satellites, little was known about the seasonal variations in ice cover, and mariners took their chances against extreme cold, harsh winds, and constantly shifting ice floes. Darkness and frequent cloud cover severely hamper observations from spacecraft, and consequently they are primarily used for large-scale observations of ice movement and extent (Fig. 7–17). Satellites with radar sensors provide all-weather, day and night observations of the polar ice cap. Radar images help identify floes and ridges in the floating ice as well as frozen leads and open water channels.

The leads, or cracks in the Arctic ice pack, affect temperatures by releasing tremendous amounts of heat into the atmosphere.

Radar altimetry data also can delineate the boundary between Arctic pack ice several feet thick and the surrounding sea. In addition, satellite radar observations can track the movement of ice floes, which travel on average about 10 miles a day. Because icebergs are an extreme hazard to shipping in Arctic waters, satellite-borne radar provides advance warning, so ships can find safer routes.

Submarine measurements have reported an unusual thinning of the Arctic pack ice north of Greenland during the winter of 1987. The thin ice was not necessarily caused by global warming, however, but was mostly due to a radical change in the pattern of ice drift. The thinning of the ice in the Arctic would be one of the first signs of global warming. A sustained global warming and possible changes in sea ice would present opportunities for increased use of the northeast and northwest passages across the Arctic Ocean. The reduction of sea ice would also decrease offshore oil-drilling hazards, but onshore development could become more difficult and expensive in regions of thawing permafrost.

LIFE BENEATH THE ICE

Surprisingly, life is abundant in the Arctic Ocean, from simple bacteria to whales. In the Arctic environment, sea ice inhibits the growth of primary producers called phytoplankton for 8 months of the year. Algae become trapped in the sea ice when it freezes from October to August. During the long, dark winter, their growth is extremely slow, but as the light returns in early spring, they proliferate rapidly at the ice-water interface.

Algae, bacteria, and other microscopic organisms begin growing on the undersides of sea ice as longer spring days bring increasing amounts of sunlight. As Arctic days become warmer, the ice breaks up and the organisms are released into the water, where they support an escalating food chain that ends with fish, seals, whales, and polar bears. Some bears live several months on the pack ice hundreds of miles from the nearest land. They subsist on abundant marine life in the open-water leads in the ice.

In the cold and dark of the Arctic winter, many species of herbivorous zooplankton like the tiny plant-eating crustaceans called copepods are important links in the marine food chain. Reproduction usually begins just before or immediately after the breakup of the ice and produces a single brood. Infant development is relatively rapid during the melting season, from April to September, with most individuals reaching adolescence before the onset of new ice.

During the winter months from October until about mid-May, virtually no phytoplankton live in the seawater. Yet in the late winter or early spring, the young copepods somehow emerge as mature adults. Apparently, the hungry copepods overwinter beneath the ice, grazing off the ice algae. Whether they can actually remove algae from under the ice or simply consume the algae melted at the ice-water interface is not known. Nevertheless, they survive the ordeals of the harsh winter and continue to make the polar oceans one of the richest environments on Earth.

If global warming melts the polar sea ice, the number of microscopic organisms at the bottom of the marine food chain would decline, and the marine animals they support would suffer as well. Less sea ice also would affect seals (Fig. 7–18), which breed on the ice-covered sea, and polar bears, which hunt and travel on the ice. Because the effects of global warming would be most extreme at higher latitudes, plants and animals living near the poles would be placed in greater jeopardy than those living near the equator.

The Arctic accounts for about 10 percent of the world's fish catch, and among the most important are cod, capelin, and herring. The Bering Sea is the world's most productive region for walleye and pollack. The pollack alone has an estimated yearly yield of over a million tons, accounting for nearly 80 percent of the foreign catch in the Bering Sea, where a single net haul can bring in 100 tons of fish. In Alaska, the salmon fishery is particularly important, especially to the natives.

Figure 7–18 A Weddell seal sunning on a pressure ridge near Scott Base, Antarctica. Photo by Luethje, courtesy of U.S. Navy

8

THE ICE CONTINENT

Antarctica is a land of ice with geographic features just like other continents, only its mountain ranges, high plateaus, lowland plains, and canyons are buried under a sheet of ice, up to 3 miles thick in some places (Fig. 8–1). The ice cap has an average thickness of about 1.3 miles and a mean elevation of about 7,500 feet above sea level. The ice is so unimaginably heavy it depresses the continental bedrock nearly 2,000 feet.

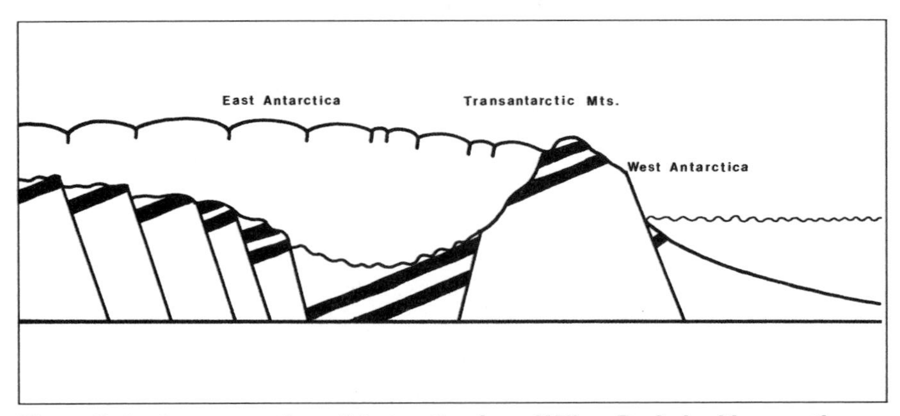

Figure 8–1 A cross section of Antarctica from Wilkes Basin looking south.

Barren mountain peaks soar 17,000 feet above the ice sheet, and winds of hurricane force with speeds up to 200 miles per hour buffet ice-laden mountains and high-ice plateaus. Yet Antarctica is literally a desert, with an average annual snowfall of less than 2 feet, which translates into only about 3 inches of rain, making the continent one of the world's most impoverished deserts. Dry valleys running between mountain ranges are the coldest and most barren places on the face of the Earth.

ANTARCTIC EXPLORATION

Greek scholars predicted the existence of Antarctica some 2,000 years before its discovery over two centuries ago, when it was stumbled upon purely by accident. The British navigator James Cook discovered *terra incognita* (unknown land) in 1774, although heavy pack ice forced him to turn back before he could make a sighting of the frozen continent. More than a century passed before explorers actually set foot on Antarctica.

Lured by Cook's reports of abundant whales and seals, American and European sealers began hunting the Antarctic waters in 1790. By 1820, the frigid waters around Antarctica were routinely visited by sealers, who slaughtered the animals by the thousands for their oil and fine pelts. A decade later, fur-seal and elephant-seal populations were all but completely wiped out.

Figure 8–2 Rough ice on the Ross Ice Shelf appears as majestic mountains and valleys. Photo by B. L. Mason, courtesy of U.S. Navy

Seldom were whales caught for their meat but mainly for their blubber, which rendered an excellent oil used for lighting and for making perfume and other products. By the turn of the century, Antarctic whales were becoming as depleted as those in the northern seas. Fisheries were also becoming exhausted, primarily by Russian trawlers.

Several countries, including the United States, Great Britain, France, and Russia, sent expeditions into the south polar seas, where explorers made the first official sightings of Antarctica. In 1839, the Scottish explorer Sir James Clark Ross attempted to locate the South Magnetic Pole. He sailed his ships through 100 miles of pack ice on the Pacific side of the continent until finally emerging into open water, known today as the Ross Sea in his honor. With his way blocked by an immense wall of ice 200 feet high and 250 miles long, known as the Ross Ice Shelf (Fig. 8–2), Ross gave up his quest to the South Magnetic Pole, about 300 miles inland of his position.

The Antarctic Ocean is the coldest marine habitat in the world and was once thought to be totally barren of life. But in 1899, a British expedition to the southernmost continent was the first to endure an entire year on Antarctica. Zoologists found examples of previously unknown fish species related to perchlike fish common throughout the world. Upward of 100 species of this fish are confined to the Antarctic region alone, accounting for about two-thirds of all fish species in the area.

In 1902, the British explorer Commander Robert Scott attempted to reach the geographical South Pole starting out from McMurdo Sound but was

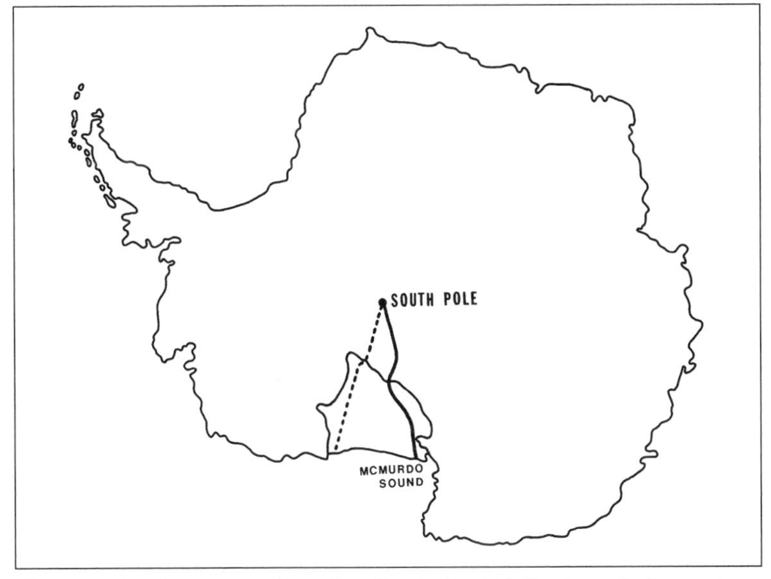

Figure 8–3 Routes taken by Scott (solid line) and Amundsen (dashed line) to the South Pole.

Figure 8–4 Admiralty Mountain Range near Hallette Station, Antarctica. Photo by G. L. Arnold, courtesy of U.S. Navy

forced to turn back after covering only about a third of the distance. In 1909, one of Scott's former team members, British explorer Ernest Shackleton, came within 97 miles of the South Pole but had to return due to low supplies and poor weather. However, one of Shackleton's teams did successfully locate the South Magnetic Pole, which in its own right was an important scientific achievement.

Scott made a second attempt to reach the South Pole in 1911, but this time he had competition from the Norwegian explorer Roald Amundsen. With his team of hardened veteran explorers, Amundsen reached the South Pole on December 15, 1911, completing the 1,600-mile round trip in less than 100 days. Scott reached the South Pole a month later, only to find that Amundsen had been there first (Fig. 8–3). On his return trip, Scott and two of his companions were caught in a raging blizzard and froze to death just 13 miles from their supply depot. Amundsen died in a tragic airplane crash in 1928 while traveling to the North Pole. Some accounts credit him for being the first to reach the North Pole as well as the South Pole.

During an expedition in the Antarctic summer of 1969-70 (which runs from December 21 to March 21), scientists discovered in the frigid cliffs of the Transantarctic Mountains, a rocky spine that transects the continent from the northwest to the southeast, a fossilized jawbone and canine tooth belonging to the mammallike reptile lystrosaurus. This unusual looking animal, measuring about 2 feet long with large down-pointing tusks, lived around 160 million years ago. The only other known fossils of lystrosaurus exist in China, India, and southern Africa. This freshwater reptile could not have swum across the salty ocean that separated the southern continents. Instead, its discovery on the frozen wastes of Antarctica was hailed as further evidence for the existence of the great southern continent Gondwana.

Protected for centuries by its forbidding climate and since June 1961 by the international Antarctic Treaty, the seventh continent with its snow-covered peaks and ice plains (Fig. 8–4) has become truly the world's last

Figure 8–5 Pole Station at the South Pole, Antarctica. Photo by H. C. Steiner, courtesy of U.S. Navy

Figure 8–6 McMurdo Station, Antarctica. Photo by Cookson, courtesy of U.S. Navy

frontier. However, in 1988, a new treaty was drawn up between the 38 nations of the Antarctic Treaty System to end the moratorium on petroleum and mineral exploration.

Concern over whether offshore drilling platforms could pose an environmental hazard in the iceberg-filled waters along the coast prompted a reassessment of the energy potential of Antarctica. A nearly 2-mile-thick, continuously flowing ice sheet covers 98 percent of the continent, offering many hazards to mining, which probably would be concentrated initially along the rocky coasts. Finally, on October 4, 1991, 31 nations signed the Madrid Protocol, banning petroleum and mineral exploration for the next 50 years.

Several nations have held territorial claims to slices of Antarctica. Russia has circled the continent with research stations, while the United States maintains a major base at the geographic South Pole (Fig. 8–5), which is the hub of every land claim. The McMurdo Station at the head of McMurdo Sound and operated by the National Science Foundation is the largest American base on Antarctica (Fig. 8–6). Its purpose is to conduct scientific research on the unusual environment of the last unspoiled continent in the world.

ANTARCTICA

Antarctica was not always shrouded in ice. About 80 million years ago, when it was still part of Gondwana, the seas around the continent were reasonably warm and only alpine glaciers covered the mountaintops. Fossils of 40-million-year-old sharks, rays, ratfish, and other marine animals now absent in Antarctica were found on Semour Island. Further evidence for a warm climate that supported lush vegetation and forests is found in coal seams running through the Transantarctic Mountains that are among the most extensive coal beds in the world. Between layers of coal are fossils of the ancient fern glossopteris found only on the southern continents that comprised Gondwana.

Near the end of the Cretaceous period, forests extended into the polar regions far beyond the present-day tree line. The most remarkable example is a well-preserved fossil forest on Alexander Island in Antarctica. To survive the harsh Antarctic conditions, the trees had to develop protection against the cold, because plants are more sensitive to the lack of heat than the absence of sunlight. Therefore, they probably adapted mechanisms for intercepting the maximum amount of sunlight when global temperatures were considerably warmer than they are today.

During the middle Cretaceous, much of Australia, which was still attached to Antarctica, wandered south of the Antarctic Circle and acquired a thick layer of ice. The interior of Australia contained a large inland sea, whereupon icebergs appear to have drifted, as indicated by huge boulders sitting in the middle of the desert that were rafted out to sea on slabs of glacial ice.

As Antarctica and Australia continued to move eastward, a giant rift began to separate them. Antarctica detached itself from South America and Australia and moved toward the south polar region. Australia continued moving into the lower latitudes, while Antarctica drifted over the South Pole, where glaciers easily grow.

Antarctica became a continent of ice about 40 million years ago, when the Earth plunged into a colder climate. The seas withdrew from the land as the ocean dropped 1,000 feet to perhaps its lowest level in the last several hundred million years. Much of the drop in sea level probably resulted from the accumulation of massive ice sheets atop Antarctica, perhaps the thickest it has ever witnessed.

The establishment of a circum-Antarctic ocean current isolated Antarctica from warm-water currents originating from the tropics, and great ice sheets spread over the eastern end of the continent. Most of the ice apparently melted during a warming trend between about 30 million and 15 million years ago. About 14 million years ago, a second major ice sheet formed as the climate grew colder. The West Antarctic ice sheet apparently formed no earlier than about 9 million years ago. Before being covered with

Figure 8–7 The archipelago of West Antarctica, shown here without its cover of ice.

ice, West Antarctica was an archipelago of scattered islands (Fig. 8–7) about the size of the Philippines.

The ice on Antarctica has fluctuated dramatically in the past few million years, vanishing completely from the entire continent at least once and from its western third perhaps several times. Supposedly, about 3 million years ago, prior to the onset of the Pleistocene ice ages, the Antarctic ice cap was virtually nonexistent. Apparently, the ice sheet had collapsed, transforming much of the continent into a cluster of islands divided by an open sea.

About 4 million years ago, great forests grew on the flanks of the Transantarctic Mountains, as evidenced by discoveries of nonfossilized wood and marine fossils. In the relatively warm climate, great open seaways appear to have reached deep into the interior of the continent, and the central ice mass might have retreated to small interior ice sheets and high alpine glaciers.

During its stay over the South Pole, Antarctica froze and thawed several times over the last 40 million years as suggested by rising and falling sea levels and fossilized plants discovered in the interior of the continent. A recent discovery of remnants of an ancient beech forest near the head of the Beardmore Glacier, approximately 250 miles from the South Pole, suggests that Antarctica was both ice free and much more temperate about 3 million years ago. Similar fossil finds made elsewhere indicate that western Antarctica was mostly ice free as recently as 100,000 years ago, around the beginning of the most recent ice age.

Most of the ice on Antarctica today accumulated during the Pleistocene. In the last ice age, the Antarctic ice volume increased about 10 percent. But during the glacial peak 18,000 years ago, the East Antarctic ice sheet actually shrunk due to the lack of precipitation. With an area of about 5.5 million square miles, this desolate world of ice is roughly twice the size of Australia. The ice sheet rises nearly 3 miles in places, with an average thickness of over 7,000 feet, amounting to about 7 million cubic miles of ice. If the Antarctic ice cap melted, sea levels could rise as much as 300 feet.

Prior to the end of the Permian period, the younger parts of West Antarctica had not yet formed, and only East Antarctica was present. East Antarctica consists of a single huge tectonic plate, composed of a Precambrian shield, whereas the West Antarctic landmass is a jumble of geologically younger small plates, called terranes, which rafted into the continent by a subducting ocean floor, and are riddled with active rifts. The Transantarctic Mountains formed along the uplifted rim of the East Antarctic shield where it meets West Antarctica. The range comprises great belts of folded rocks, upraised when two plates came together into the continent of Antarctica.

West Antarctica is dominated by the Ross and Weddell seas, which are covered by thick, floating ice shelves. The elevation is quite low, with most of the ice resting on ground comprised of glacial till much of which is below sea level. The till is a mixture of ground-up rock and water that acts as a lubricant, allowing the ice sheet to slide into the ocean.

The Antarctic Filchner-Ronne Ice Shelf, located south of the Weddell Sea, is the most voluminous on Earth. It consists of two distinctive layers. The upper layer measures about 500 feet thick and consists mostly of ice formed by falling snow. The bottom layer measures about 200 feet thick and consists of frozen seawater. The freshwater layer consists of opaque and granular ice resembling the top portion of a glacier. In contrast, the transparent marine shelf ice displays many inclusions of marine origin such as diatoms, radiolarians, and clay particles. Free-floating ice platelets recrystallize at the base of the marine layer, forming a slush that slowly compacts into solid ice.

The ice in East Antarctica is fairly stable. It is so thick the region is practically devoid of earthquakes because the massive ice sheets stabilize earthquake faults and inhibit fault slip. However, Antarctica is noted for its many volcanoes, the most active of which is Mount Erebus (Fig. 8–8), a smoldering mountain that rises 12,500 feet above Ross Island. It was visited in early 1993 by the eight-legged Dante robot, which took a perilous journey down the volcanic vent and collected rock and gas samples. Several other

Figure 8–8 Mount Erebus, Ross Island, the major active volcano in Antarctica.
Photo by D. R. Thompson, courtesy of U.S. Navy

volcanoes poke through the ice of West Antarctica across the Ross Ice Shelf from the Transantarctic Mountains about 900 miles from the South Pole.

Many of Antarctica's dormant volcanoes are buried within the ice, and extensive volcanic deposits underlie the ice sheets. When active volcanoes erupt underneath the ice, they produce great floods of meltwater. The meltwater mixes with the underlying sediment, forming glacial till tens of feet thick. Basalt erupted beneath the glacial ice produce hyaloclastics, which are pillow lavas and pillow breccias that are quick-frozen forms of lava.

Despite all the ice on Antarctica, it is literally a desert. Dry valleys gouged out by local ice sheets running between McMurdo Sound and the Transantarctic Mountains (Fig. 8–9) receive less than 4 inches of snowfall each year, most of which is blown away by strong winds. Some regions have not received precipitation for up to a million years.

By a freak of nature, Antarctica happens to be the best hunting ground for meteorites, which stand out in stark contrast to the white snow and ice. Meteorites landing on the ice over the years tend to concentrate when the glaciers melt or sublimate, which is a reduction in ice volume by evaporation without melting. A few rare meteorites found on the ice sheets of

Antarctica might be parts of the Martian crust blasted out by large asteroid impacts millions of years ago and are just now reaching the Earth. Even pieces of the moon appear to have landed on the ice surface when major asteroid impacts blasted them into space and hurled them at the Earth.

Figure 8–9 Wright Dry Valley, Antarctica. Photo by D. R. Thompson, courtesy of U.S. Navy

Figure 8–10 A portable automatic weather station mounted on a sled and used to transmit weather data to McMurdo Station, Antarctica. Photo by J. Horton, courtesy of U.S. Navy

Snow falling on Antarctica accumulates into thick ice sheets because virtually no melting occurs from year to year. The summer mean monthly temperature at the South Pole is about -30 degrees Celsius. The winter mean monthly temperature is -60 degrees, and in places the temperature has been known to drop to -90 degrees. Only automated remote weather stations placed strategically across the continent (Fig. 8–10) can tolerate such frigid conditions, which are almost as cold as on the surface of Mars.

THE ICE STREAMS

Nine-tenths of all ice in the world buries Antarctica, which contains 70 percent of all the Earth's fresh water. The continent is divided by the

Transantarctic Range, a wall of mountains that forms the spine of Antarctica and separates the eastern and western ice sheets into a large East Antarctic ice mass and a smaller West Antarctic lobe about the size of Greenland. The ice sheet in East Antarctica is firmly anchored on land. But the ice sheet in West Antarctic rests below the sea on bedrock and glacial till and is surrounded by floating ice pinned in by small islands buried below the ice.

Ice streams several miles broad traverse West Antarctica, and rivers of solid ice flow down mountain valleys to the sea. The ice streams flow through stationary or slower-moving glaciers that are bounded on each side by rock. They carry more ice to the coast than ice accumulates at the stream sources. Some ice streams travel quite rapidly. For example, the Rutford Ice Stream cruises along at a speed of over 1,200 feet a year, roughly 100 times faster than the ice bordering it.

Rivers of ice slowly flow outward and down to the sea on all sides of the Transantarctic Range. The streams of glacial ice flow down from the mountains and onto giant floating ice shelves. The ice streams act like pipelines that carry ice from the stable continental interior toward the ocean, where they form floating ice shelves that break apart into icebergs. Muddy pools of meltwater lubricate the base of the ice streams, allowing them to glide smoothly along the valley floors.

The ice escapes through mountain valleys to the ice-submerged archipelago of West Antarctica, and onto the ice shelves of the Ross and Weddell Seas (Fig. 8–11). During the height of the last ice age 18,000 years ago, the

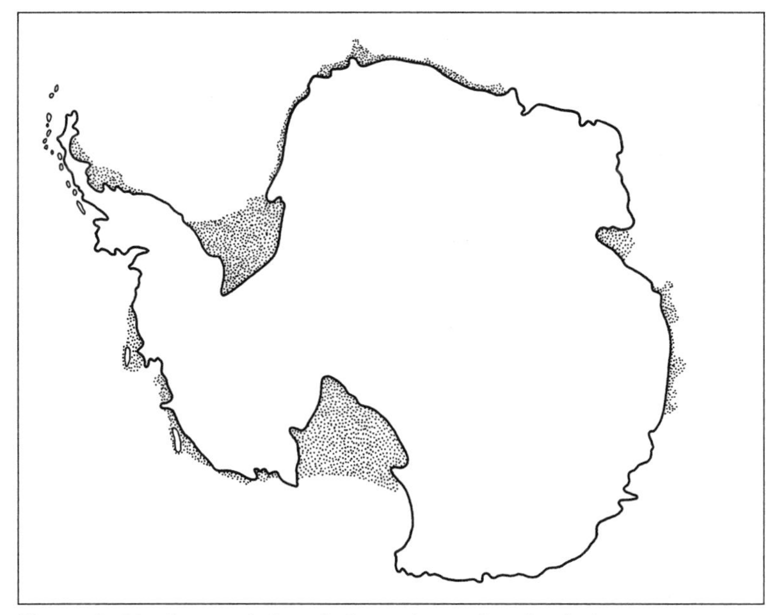

Figure 8–11 Antarctic ice shelves shown in stippled areas.

Ross Ice Shelf slumped to the seafloor, causing the ice streams feeding it to back up. The ice shelf greatly expanded and flowed upward into the surrounding valleys and buried them.

Normally, ice shatters when placed under stress. But because of its large size, a glacier acts like a flowing viscous solid, creeping over the landscape at approximately half a mile per year. Friction at the base of the glacier is lost due to huge subglacial lakes and streams. When a glacier surges, the streams stop flowing and the water spreads out beneath the glacier. This watery undercoat acts as a lubricant, and large parts of the ice sheet surges along the ice streams toward the sea at speeds several times faster than usual.

The glaciers glide along on the valley floor and eventually plunge into the sea. The amount of ice flowing to the coast is significantly greater than the quantity accumulating at the ice stream's source, indicating a possible instability that warrants further study to understand the nature of ice streams in order to predict their future behavior.

The ice on Antarctica is not uniform but pervaded by internal layers. Large, flat areas beneath the ice are thought to be subglacial lakes, kept from

Figure 8–12 The heavily crevassed surface of the Columbia Glacier.
Courtesy of USGS

Figure 8–13 A large iceberg along the coast of Grahm Land, Antarctic Peninsula.
Photo by W. R. Curtsinger, courtesy of U.S. Navy

freezing by the Earth's interior heat. The temperature a mile down can be 25 degrees warmer than on the surface of the ice. Because of the high pressures at such depths, liquid water can exist several degrees colder than its normal freezing point. The pools of liquid water tend to lubricate the ice streams and help them flow down the mountain valleys to the sea, allowing ice streams up to several miles broad to glide smoothly along the valley floors.

The interior portions and banks of the ice streams are marked by deep crevasses. Glacial crevasses are cracks or fissures in a glacier, resulting from stress due to movement in the ice (Fig. 8–12). They are generally several tens of feet wide, a hundred or more feet deep, and up to a thousand or more feet long. Deep crevasses often flank the banks of glaciers, where they make contact with the walls of the glacial valley.

Crevasses also run parallel to each other down the entire length of the ice streams, especially when the central portion of the glacier flows faster than the outer edges. Snow bridges occasionally span the crevasses, in some cases completely hiding them from view. Sometimes a stream of meltwater can be heard gurgling far below from open crevasses slicing through the glacier.

Antarctica discharges over a trillion tons of ice into the surrounding seas annually, and the ice calves off into icebergs (Fig. 8–13). The icebergs also appear to be getting larger, and the number of extremely large icebergs has also increased dramatically due possibly to the warming of the Earth's climate. Antarctica has several ice shelves, which are thick sheets of floating, fresh-water ice that have slid off the continent. Thousands of years of snow accumulation forms the continental ice cap, which flows down to the sea, where it replenishes the parts of the ice sheet that break off and head for the open sea.

The West Antarctic ice sheet is inherently unstable, and during a sustained warmer climate it could suddenly collapse, and the rapidly melting, unstable ice sheet would break loose and crash into the sea. The additional sea ice would raise global sea levels upward of 20 feet and inundate coastal areas. The ice sheet has collapsed into the ocean at three irregular intervals of 750,000, 330,000, and 190,000 years ago, which did not necessarily correspond with periods of global warming.

Even a slow melting of both polar ice caps could raise the level of the ocean up to 12 feet by the end of the 21st century and drown much of the world's coastal plains and flood coastal cities. A rise in sea level also would lift West Antarctica ice shelves off the seafloor and set them adrift into warm equatorial waters, where they would rapidly melt, further raising the sea.

As the present climate continues to warm, it could produce an instability in the Antarctic ice sheets, causing them to calve off into the ocean and make additional sea ice. The increased area of ice in the Antarctic Ocean could form a gigantic ice shelf, covering as much as 10 million square miles. The additional ice would increase the Earth's albedo, which would lower temperatures in the Southern Hemisphere.

THE CIRCUM-ANTARCTIC CURRENT

When Antarctica separated from South America and Australia and moved over the South Pole about 40 million years ago, a circumpolar Antarctic ocean current encircled the continent between 50 and 60 degrees south latitude (Fig. 8–14). The circum-Antarctic current isolated Antarctica from the rest of the ocean, preventing the continent from receiving warm poleward flowing waters from the tropics, resulting in a frozen wasteland.

As the continent continued to cool, it became covered by a thick layer of ice, which even dwarfed the present ice sheet. All land features including canyons, valleys, plains, plateaus, and mountains were buried beneath the ice. Thick glaciers buried even the highest mountains under at least 1,500 feet of ice, which extended across the sea as far as the tip of South America.

When Antarctica was connected to a larger landmass, it received warm ocean currents originating from the tropics, which kept it relatively ice free. However, when it broke away from Gondwana during the Cretaceous and drifted over the South Pole, a new ocean current was established. Easterly

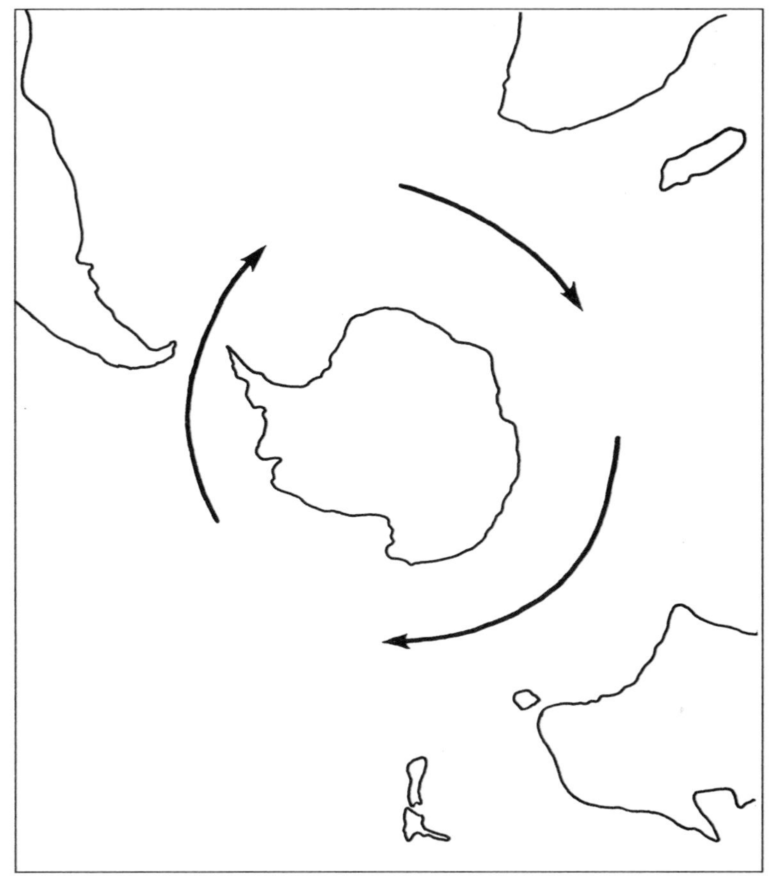

Figure 8–14 The circum-Antarctic current isolates the continent from warm water currents.

blowing polar winds circling the South Pole pushed against the sea as they blew across the ocean's surface. The friction between the wind and the waves it generated set in motion a permanent ocean current that surrounded the continent, flowing around it like a snake chasing its tail.

Around 30 million years ago, the largest of the ice caps expanded across the continent. Sometime during the next 15 million years, most of the ice sheet melted, probably due to a warmer global climate. Some 14 million years ago, a new ice cap formed in its place as the climate grew colder and the ocean's bottom temperature approached the freezing point of water (Fig. 8–15).

Most of the ocean's densest water comes from Antarctica's Weddell Sea, where cold air and the exclusion of salt from freezing seawater combine to send surface water to the bottom of all the major oceans. This Antarctic Bottom Water, which is the major mass of cold, salty bottom water in the world, flows northward well beyond the equator.

The ocean is filled with icy water originating at the polar seas. Most ocean water is within a few degrees of freezing, with the only exception being a thin layer of warm surface water mostly in the tropics. The buoyancy of seawater gives the ocean a layered structure, and masses of water are stratified according to their density as determined by their temperature and salinity. Mixing between layers is restricted because water density increases with depth. Instead of crossing density surfaces, water prefers to travel along the plane of a single layer.

The Antarctic Ocean is a pivotal part of an enormous heat engine that drives the motions of much of the world's ocean. Furthermore, the massive overturning of water in the Antarctic dramatically affects the chemistry of the deep ocean, replenishing the waters with oxygen depleted by living organisms. The deep-ocean overturning also might play a significant role by balancing the oceanic concentrations of dissolved gases such as carbon dioxide with the level of those gases in the atmosphere, with important implications for the greenhouse effect.

About 57 million years ago, ancient Antarctica, which was much warmer than today, generated cool water that filled the upper layers of the ocean,

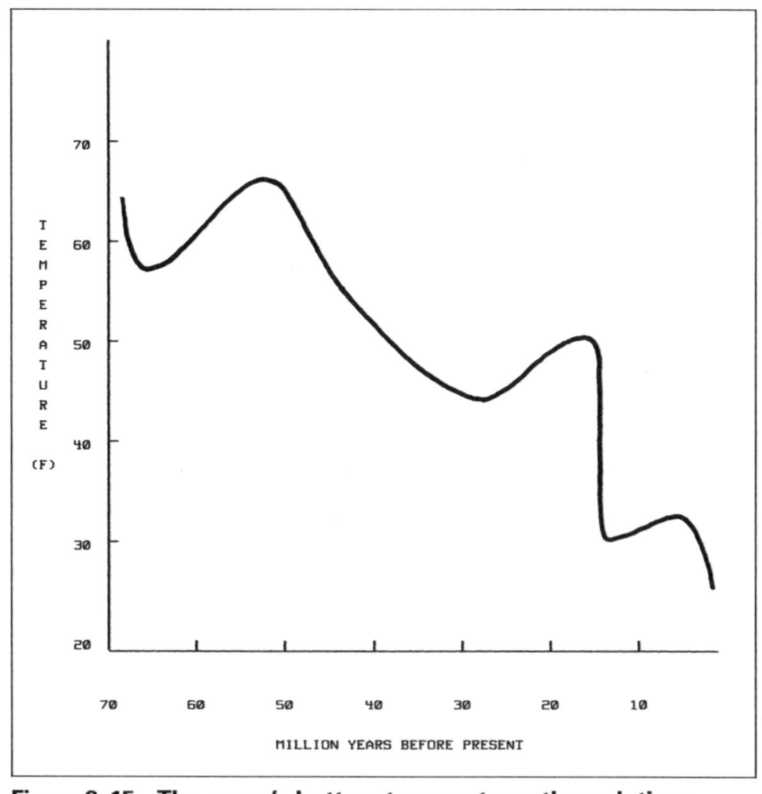

Figure 8–15 The ocean's bottom temperatures through time.

TABLE 8–1 HISTORY OF DEEP CIRCULATION IN THE OCEAN

Age (million years ago)	Event
3	An ice age overwhelms the Northern Hemisphere.
3–5	Artic glaciation begins.
15	The Drake Passage is fully opened; the circum-Antarctic current is formed. Major sea ice forms around Antarctica, which is glaciated, making it the first major glaciation of the Modern Ice Age. The Antarctic bottom water forms. The snow limit rises.
25	The Drake Passage between South America and Antarctica begins to open.
25–35	A stable situation exists with possible partial circulation around Antarctica. The equatorial circulation is interrupted between the Mediterranean Sea and the Far East.
35–40	The equatorial seaway begins to close. There is a sharp cooling of the surface and of the deep water in the south. The Antarctic glaciers reach the sea with glacial debris in the sea. The seaway between Australia and Antarctica opens. Cooler bottom water flows north and flushes the ocean. The snow limit drops sharply.
>50	The ocean flows freely around the world at the equator. Rather uniform climate and warm ocean exists even near the poles. Deep water in the ocean is much warmer than it is today. Antarctica only has alpine glaciers, with no major ice sheets.

causing the entire ocean circulation system to run backward. Instead of flowing from the poles to the deep water and rising to the tropical surface waters, the global circulation ran in the opposite direction, with numerous implications for climate and life.

Around 28 million years ago, Africa collided with Eurasia, turning off the tap of warm water flowing to the poles, placing Antarctica in the icy grip of glaciation. The cold air and ice cooled the surface waters and made them sufficiently heavy to sink and flow toward the equator, providing the oceans with the circulation system it has today.

Figure 8–16 Marine life on the bottom of McMurdo Sound, Antarctica. Photo by W. R. Curtsinger, courtesy of U.S. Navy

ANTARCTIC LIFE

No trees, bushes, or a single blade of grass grow on the entire Antarctic continent. Only meager signs of life exist, consisting of blue-green algae on the bottom of small glacier-fed lakes, soil bacteria, and a giant wingless fly. Only two flowering plant species live on Antarctica, which have undergone population explosions, possibly due to a warming climate.

Delicate mosses and lichens if disturbed take a century to grow back. The discovery of lichens in the tiny pores on the undersides of some Antarctic rocks has even fueled speculation that life forms might inhabit Mars. Indeed Antarctica is an ideal place to train for preparations for future excursions to the red planet.

The Antarctic Ocean is the coldest marine environment in the world, yet the seas surrounding Antarctica are teeming with life (Fig. 8–16). The waters around Antarctica are cut off from the general circulation of the ocean by the circum-Antarctic current, which provides a thermal barrier that impedes the inflow of warm currents and tropical fish as well as the outflow of Antarctic fish. Moreover, due to the extreme cold and low productivity, the Antarctic Ocean has less diversity than the Arctic Ocean,

Figure 8–17a A U.S. Coast Guard icebreaker clearing a path through the ice in McMurdo Sound, Antarctica. Photo by T. F. Ahlgrim, courtesy of U.S. Navy

which supports almost twice as many species. Nevertheless, the frigid waters of the Antarctic Ocean are home to species of birds and mammals found nowhere else on Earth.

During the Antarctic winter, from June to September, sea ice covers nearly 8 million square miles of ocean that surrounds the continent (Fig. 8–17a and b), an area that is over twice the size of the United States. The sea ice is punctured in various places by coastal and ocean polynyas, which are large open-water areas kept from freezing by upwelling warm-water currents. The coastal polynyas are like sea ice factories because they expose portions of open ocean that later freeze, continuing the ice-making process. These open bodies of water release a tremendous amount of heat into the atmosphere, which can rise high into the stratosphere. The upwelling water also has a high carbon dioxide content and releases large quantities of this gas into the atmosphere.

The Antarctic Ocean is the planet's largest coherent ecosystem, representing about 10 percent of the total extent of the world's ocean. It teems with life from microscopic phytoplankton (one-celled floating plants) to killer whales and leopard seals. Whales, seals, fish, squid, and seabirds prey on abundant tiny shrimplike crustaceans called krill, which in turn feed on phytoplankton at the very bottom of the food chain.

As whale populations decline, other krill-eating animals have shown rapid population increases in recent years, causing a subsequent decline in krill. Population increases in Antarctic seals have outpaced any simple recovery from past overhunting. Some seabird population increases have been documented as well. Similarly, penguin populations (Fig. 8–18) are unexplainably larger after the slaughter of the 19th century. The penguin is one of the world's hardiest birds, able to nest along the harsh Antarctic coastline.

Four months out of the year Antarctica is in total darkness, and sea ice up to 10 feet or more thick covers the water at least 10 months. Even during the short summer season, the water under the ice receives less than 1 percent of the sunlight on the surface. The water temperature throughout the year varies from about 1 to 2 degrees below freezing, but due to its high salt content the water is kept in a liquid state.

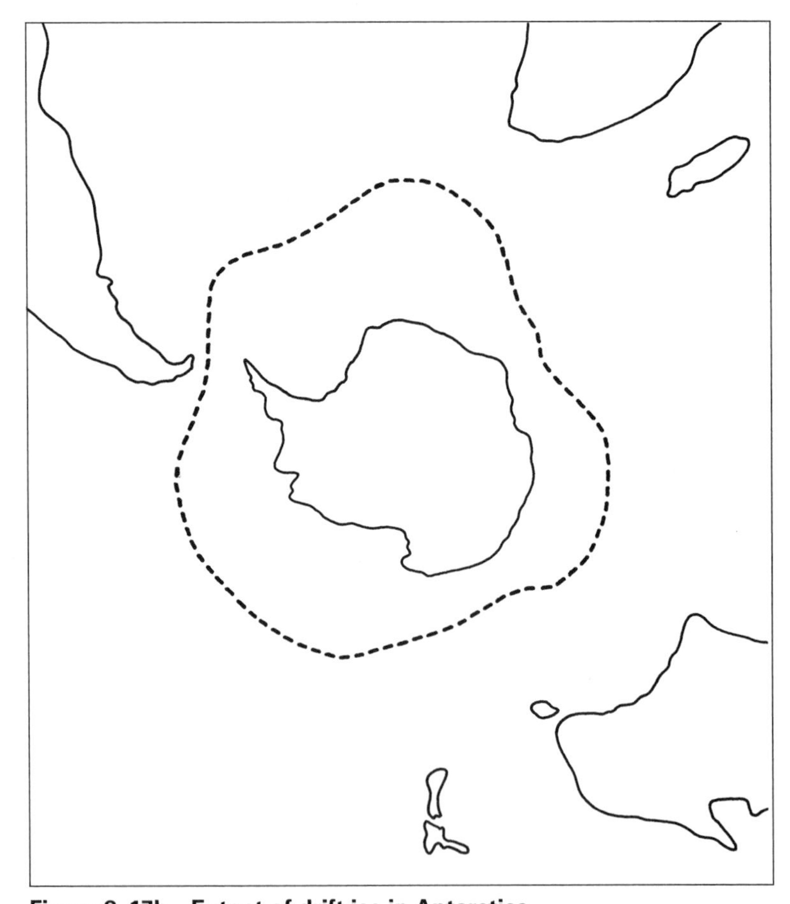

Figure 8–17b Extent of drift ice in Antarctica.

Figure 8–18 A pair of Adélie penguins in Antarctica. Photo by M. Mullen, courtesy of U.S. Navy

Because of these extreme conditions, the Antarctic Ocean has only about half the number of species as its northern counterpart. Yet although seawater remains a few degrees below freezing throughout the year, certain species of fish thrive beneath the ice. This is because special proteins in their bodily fluids act as an antifreezelike substance, which inhibits the propagation of ice crystals in their bodies that would damage tissue and cause death. Because fish are cold-blooded, their body temperature is essentially the same as their environment. Therefore, in order to survive the long, dark Antarctic winters in freezing waters with a scarce food supply, fish must adapt to special conditions found nowhere else on Earth.

9

GLACIAL STRUCTURES

During the last ice age, many northern regions were sculpted by immense glaciers that swept down from the polar regions. Continental glaciers enveloped entire continents and alpine glaciers grew on mountain peaks that are now ice free. Their legacy remains as carved-up rock in the high ranges and plains around the world.

Glaciers are perhaps the most effective erosional agents. This effect is manifested by steep-sided valleys carved out by massive rivers of ice flowing out of the mountains. The glaciers left behind a bizarre collection of structures, including cirques, potholes, kettles, glacial lakes, flood-ruptured ground, and many other landforms carved out by ice.

GLACIAL EROSION

Most ancient geologic structures have long been erased by the Earth's active erosional processes, including the action of wind, rain, glacial ice, freezing and thawing, and plant and animal life. The hydrologic cycle, involving the flow of water from the ocean to the land and back to the sea (Fig. 9–1), provides powerful erosional forces. Areas lacking rainfall or snowfall, such as deserts and tundra regions, retain much of their geologic structures through time.

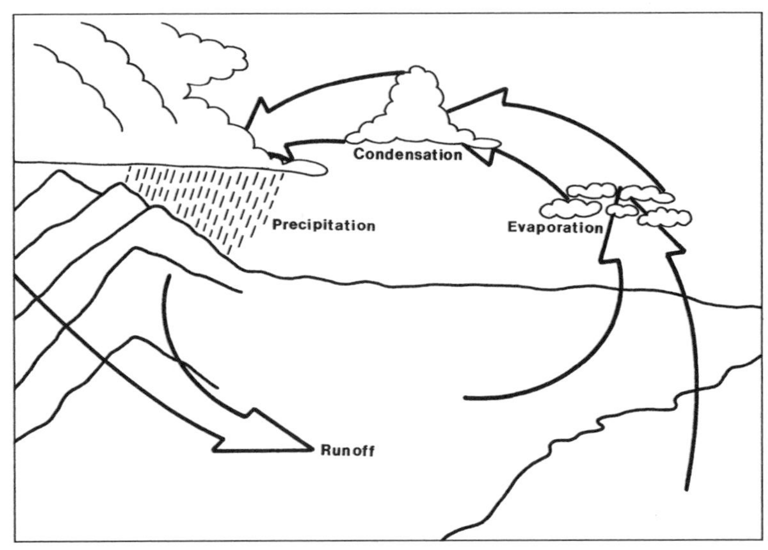

Figure 9–1 The hydrologic cycle involves the evaporation of water from the ocean and continents, the condensation into clouds, the precipitation on the land, and the return of runoff into the sea.

Figure 9–2 Chickamin Glacier is a composite valley and slope glacier, Glacier Peak Wilderness, Skagit County, Washington. Photo by A. Post, courtesy of USGS

Erosion is a geologic process that levels the tallest mountains, gouges deep ravines into the hardest rock, and obliterates most geologic features on Earth. No matter how pervasive is the formation of mountain ranges by the forces of uplift, they eventually lose the battle with erosion and are worn down to the level of the prevailing plain, with only their deep roots remaining to mark their existence. Massive glaciers carved out of the Earth some of the most monumental landforms. Outwash streams from glacial meltwater also played a major role in sculpting the landscape.

Glaciers are among the most effective agents of erosion, especially in mountainous regions, where they have the advantage of exposed rock upon which to work (Fig. 9–2). Glacial erosion causes a reduction of the land surface by processes involving glacial ice as an agent, including the transportation of rock fragments by glaciers, bedrock scouring, and the erosive action of meltwater streams.

Glacial erosion radically modifies the shape of stream valleys occupied by glaciers. Most of the erosion involves the removal of rock by plucking, and abrasion generally smooths and polishes the resulting form. Any small hills or knobs in the valley overridden by the glacier are rounded and smoothed by abrasion. Glacial erosion is most active near the head of a

Figure 9–3 A U-shaped glaciated valley at Red Mountain Pass south of Ouray, Ouray County, Colorado. Photo by L. C. Huff, courtesy of USGS

glacier, where the ice deepens and flattens the gradient of the valley.

The melting of the great ice sheets at the end of the last ice age sent massive floods raging across the land, as water gushed from trapped reservoirs beneath the glaciers. While flowing under the ice, water surged in vast turbulent sheets that scoured deep grooves in the crust, forming steep ridges carved out of solid bedrock. Huge torrents of meltwater laden with sediment surged along the Mississippi River to the Gulf of Mexico, widening its channel considerably. Many other rivers overreached their banks to carve out new floodplains.

The most recent glacial period is the best studied of all because evidence of each preceding ice age was erased by the last one, as ice sheets eradicated much of the northern landscape. The power of glacial erosion is well demonstrated by deep-sided valleys carved out of mountain slopes (Fig. 9–3) by flowing ice a mile or more thick. The glaciers descended from the mountains and spread across most of the northern lands, destroying everything in their paths.

The basement rocks that formed the nuclei of the continents are exposed in broad, low-lying, domelike structures called shields. The shields are extensive uplifted areas surrounded by sediment-covered bedrock called continental platforms, which are broad, shallow depressions of basement complex filled with nearly flat-lying sedimentary rocks. Many shields like the Canadian Shield, which covers most of eastern Canada, are fully exposed in areas that were ground down by flowing ice sheets during the last ice age. The soils in these regions are thin from glacial erosion and could soon wear out with extensive agriculture.

Figure 9–4 Polished and striated rocks near Cathedral Lake, Yosemite National Park, California. Photo by G. K. Gilbert, courtesy of USGS

Figure 9–5 A glaciated granite cirque with glacier frozen lake in the foreground, Big Horn Mountains north of Cloud Peak, Big Horn County, Wyoming. Photo by N. H. Darton, courtesy of USGS

EROSIONAL FEATURES

A glacier abrades the bedrock over which it moves by glacial scouring from the action of grinding or rasping. The abrasive agent is the rock material dragged along by the glacial ice. The ice itself, because it is much softer, does not erode the rock. Instead, it plucks fragments of bedrock by the plastic flow of the ice around them, and the rocks become part of the moving glacier. Boulders embedded in the ice gouge deep cuts into the easily eroded bedrock. Smaller rocks produce striations or scratches in the bedrock, while finer material polishes it to a smooth finish (Fig. 9–4).

These erosional features led geologists to the discovery of other historic ice ages. The continents of South Africa, South America, Australia, Antarctica, and India show evidence of contemporaneous glaciation during the late Paleozoic, around 270 million years ago, as indicated by deposits of glacial till and grooves in the ancient rocks excavated by boulders encased in slowly moving masses of ice. The scars were created by rocks embedded at the base of glaciers that scoured the landscape.

The Coteau des Prairies is a 200-mile-long delta-shaped landform in eastern South Dakota. The low-relief formation is a deposit of hard quartzite (metamorphosed quartz sandstone) that split the southward-flowing glacier of the last ice age into two lobes. The ice scoured the lowlands on either side but never covered the Coteau itself, leaving it standing alone above the adjacent terrain.

Glaciers in the alpine regions flowed down mountain peaks and gouged large pits with an amphitheaterlike form called cirques (Fig. 9–5), from the French word for "circle." They are semicircular basins or indentations with steep walls high on a mountain slope at the head of a valley and are associated with the erosive activity of a mountain glacier. Cirque walls are cut back by the disintegration of the rocks lower down the mountainside. When the glacier melts, the rock material embedded in the glacier gouges a concave floor that might contain a small mountain lake called a tarn.

The expansion of adjacent cirques produces arêtes, horns, and cols. An arête, from the French word for "fish bone," is a sharp-crested, serrated, or knife-edged ridge separating the heads of abutting cirques that once contained alpine glaciers. It also forms a dividing ridge between two parallel valley glaciers. The Continental Divide, which is the boundary between river systems flowing toward opposite sides of the continent, follows an imposing arête in Glacier National Park, Montana, known as the Garden Wall.

Three or more cirques eroding toward the same point form a triangular pyramid-shaped peak called a horn. The Matterhorn in the Swiss Alps is a perfect example of a glacial horn. A col (Fig. 9–6) is a sharp-edged or

Figure 9–6 A glaciated col between Mount Huxley and Mount Spencer, Kings Canyon National Park, Fresno County, California. Photo by G. K. Gilbert, courtesy of USGS

Figure 9–7 Small potholes on the surface of a large rock mass, Warrior River, Tuscaloosa County, Alabama. Photo by C. Butts, courtesy of USGS

saddle-shaped pass in a mountain range formed by the headward erosion where cirques meet or intercept each other.

Potholes are a common sight on exposed ancient river bottoms (Fig. 9–7). Those described as "glacial" are often found in the northern regions. The glaciers do not actually cut potholes themselves, however, but release vast quantities of water when they melt, and river channels overflow with meltwater that causes severe erosion and potholes. Another indirect effect glaciers have on pothole formation occurs during the drainage of large lakes fed by glacial meltwater. The gradients of streams flowing into the lakes greatly steepen, during which they erode downward cutting potholes.

Most potholes are smooth-sided circular or elliptical holes in hard bedrock such as granite or gneiss, which are coarse-grained igneous and metamorphic rocks. They have similar shapes, but vary in size, with diameters and depths of up to 5 feet and more. Huge, rounded boulders lying on the bottom were responsible for creating the potholes, as torrents of water from a melting ice age glacier whirled the rocks around in the holes, widening while simultaneously deepening them. Most potholes form where water is swiftly flowing and turbulent, such as streams with steep gradients and irregular beds or where a large volume of water is forced through a restricted channel when glaciers melt.

GLACIAL VALLEYS

During the last ice age, glaciers buried many mountains in Northern Europe and North America, where rivers of ice linked the Rocky Mountains with ranges in northern Mexico. Small ice sheets expanded in the mountains of Australia, New Zealand, and the Andes of South America. Throughout the world, alpine glaciers grew on mountains that are presently ice free. Their legacy remains as steep-sided valleys sculpted by roving glaciers.

Symmetrically formed glacial valleys carved out of mountain slopes by thick sheets of flowing ice well demonstrate the power of glacial erosion (Fig. 9–8). A glacial valley is a stream valley that was glaciated during the ice ages. Glaciers did not cut the original valley, however, but only modified or changed the shape of the existing one. The glaciers converted the formerly V-shaped valley into a U-shaped one with a broad, flat bottom as much as a thousand or more feet deep.

A glacier moving down a valley tends to straighten it because glacial ice cannot turn as abruptly as could the original river due to its greater viscosity. This action removes projecting spurs, which are ridges extending laterally from a mountain range, and other ridges on the insides of curves of the stream valley. The floor of the glaciated valley tends to be irregular because ice erodes more readily in fractured or weak rock, resulting in differential erosion that formed giant steps at intervals along the length of the valley.

Figure 9–8 Saskatchewan Glacier, showing eroded glacial valley, Alberta, Canada. Photo by H. E. Malde, courtesy of USGS

Figure 9–9 The Bridal Veil Falls, Yosemite National Park, Mariposa County, California. Photo by F. E. Matthes, courtesy of USGS

The volume of ice in the main-stream valley, whose source is near the crest of the mountain range, is generally much greater than the amount of ice in a tributary stream valley. Therefore, the ice eroded the main valley deeper than the tributary. After the ice melts, the tributary stream flows through a hanging valley above the main stream, into which it pours from a waterfall such as the magnificent falls in Yosemite National Park, California (Fig. 9–9).

Ice a mile or more thick buried many valleys during the ice ages. The glaciers descended from the mountains, spread across most of the northern lands, and like huge bulldozers they destroyed everything in their paths.

As the glaciers extended far down the valleys, they ground rocks on the valley floors as the ice advanced and receded. In effect, a river of solid ice embedded with rocks moved along the valley floors, grinding them down like a giant file as the glacier flowed back and forth over the bedrock. The advancing glaciers carved parallel furrows or striations in the valley floors as they sliced down mountainsides.

Downstream from the foot of a glacier, the surfaces of projecting rocks along the glacial valley floor are noticeably different from those high up on the sides of the valley. Rocks higher up are rough and jagged compared with those lower down, which are rounded, smooth, and covered with numerous parallel grooves pointing down the valley. Miles from existing glaciers are large areas of polished and deeply furrowed rocks, and rock heaps marked the extent of former glaciers.

Glacial striae are fine-cut parallel or nearly parallel lines on a bedrock surface (Fig. 9–10), inscribed by the overriding ice. They are small grooves or scratches cut into the bedrock by rock fragments carved by a glacier or cut on the transported rocks themselves and are excellent indicators of the direction of glacial flow. Glacial striae cover large areas of northern North America and Europe. They usually associate with Pleistocene deposits but

Figure 9–10 A glaciated ledge showing intensive striae marks the crest of the slope of a preglacial valley of the Nashua River, Worcester County, Massachusetts. Photo by W. C. Alden, courtesy of USGS

also occur on rocks that have been glaciated at earlier periods as far back as the Precambrian era.

The Transantarctic Range was uplifted when crustal plates comprising Antarctica collided. It divides the continent between East Antarctica and smaller West Antarctica. Glaciers heading toward the sea gouged out deep dry valleys running between McMurdo Sound and the Transantarctic Mountains. Valleys once occupied by glaciers during the last ice age are now among the most barren places on Earth, completely devoid of any life. Interestingly, experiments to determine the existence of life on Mars were conducted in the dry valleys because the two landscapes have so much in common.

A tidewater glacier on the coast can erode its valley floor below sea level. When the glacier melts, a steep-walled, troughlike arm of the sea results in a fjord (Fig. 9–11). During the ice ages, glaciers gouged fjords out of coastal mountains in Norway, Greenland, Alaska, British Columbia, Patagonia, and Antarctica.

Fjords are long, narrow, steep-sided inlets in mountainous glaciated coasts. They occur where the sea invades deeply excavated glacial troughs after the glaciers have melted. The side walls are characterized by hanging valleys and tall waterfalls. In Norway, terminal moraines produce small deltas that provided the only available flat land. Large waves generated by frequent rockfalls into the fjords can cause considerable damage as they burst through local villages.

Figure 9–11 Wolstenholme Fjord looking toward Dike Mountain, Nunatarssuaq region, Greenland. Photo by R. B. Colton, courtesy of USGS

Figure 9–12 A kettle hole in gravels near the terminus of Baird Glacier, Thomas Bay, Petersburg district, southeastern Alaska. Photo by A. F. Buddington, courtesy of USGS

GLACIAL LAKES

Many of the northern lands are dotted with glacial lakes excavated by moving glaciers and filled with water. Glacial lakes over much of Canada and northern United States resulted when glaciers eroded large depressions into the bedrock. Remote glacial lakes and tundra in northwest Canada hosts some of the oldest rocks in the world, dating to about 4 billion years ago.

Lake Agassiz at the edge of the retreating ice sheet in south Manitoba, Canada, was a vast reservoir of glacial meltwater much larger than any of the existing Great Lakes. It formed in a huge bedrock depression carved out by great sheets of ice. Similar large reservoirs of meltwater include Lake Lahonton, west of the Rocky Mountains, and Lake Bonneville in Utah and Nevada.

Smaller lakes formed when large blocks of ice buried by glacial outwash sediments melted, forming deep pits called kettles (Fig. 9–12). The depressions are circular or elliptical because ice blocks tend toward roundness when they melt. If a block of ice remains deeply buried, it is called a "glacière," an underground ice formation that stays frozen year-round even in temperate climates. It can form a cavity in which an ice mass remains unthawed throughout the year.

Kettles range up to 10 or more miles in diameter and up to 100 or more feet deep. They might occur singly or in groups (Fig. 9–13). When large numbers group together, the terrain appears as basins and mounds, called knob and kettle topography. It is an irregular assemblage of knolls, mounds, or ridges between depressions or kettles that might contain swamps or

Figure 9–13 Numerous kettles stud the tundra of Cape Bathurst and the Mackenzie River area, Northwest Territories, Canada. Courtesy of NASA

ponds. Not all kettles hold water, and some dry kettles might possess a stand of trees, which descend toward the center of the kettle. The undulating landscape is a type of terminal moraine, possibly resulting from slight oscillations of an ice front as it recedes. A section of knob and kettle topography that might have developed either along a live ice front or around masses of stagnant ice is called hummocky moraine.

The Great Lakes bordering between the United States and Canada are the largest of the glacial lakes. However, in terms of depth and volume, Russia's Lake Baikal is the largest freshwater lake, holding one-fifth of all the fresh water in the world, more than all the Great Lakes combined. Huge quantities of sediment derived from the continent flow into the lakes, and the constant buildup gradually shallows them. Eventually, the lakes will completely fill with sediment and became dry, flat, featureless plains, until the next ice age when the glaciers return to scour out their basins.

The Great Lakes possess their own unique weather systems. A seiche is a wave caused by a sudden change in barometric pressure on a large lake or bay where the water tends to slosh back and forth. They are common on

Lake Michigan, and, on occasions, they can be quite destructive to shorelines. The world's busiest waterway was also hit by an unprecedented hurricane that destroyed some 20 ships in 1913. Lake-effect snowstorms are commonplace for regions east of the Great Lakes during winter.

The rising waters of the Great Lakes threaten beaches and coastal houses. Toxic pollutants, including DDT and PCBs, rain directly into the lakes or run off contaminated areas on shore. Mussels and zooplankton concentrate the pollutants in the food chain, which ultimately taints other species. These toxins are the leading environmental issue for the Great Lakes because upward of 100 years or more are required for their polluted waters to drain into the Atlantic Ocean.

GLACIER BURST

While the continental glaciers melted, massive floods flowed across the land. Perhaps the greatest flood ever to wash over the Earth occurred deep in the Altai Mountains of southern Siberia around 14,000 years ago, during the melting of the great ice sheets. Near the end of the last ice age, a glacier moved out of a valley perpendicular to the Chuja Valley and cut across it, damming the valley and creating a large lake nearly 3,000 feet deep that held some 200 cubic miles of water.

Figure 9–14 The Channeled Scablands of the northern end of the upper Grand Coulee River, eastern Washington. Photo by F. O. Jones, courtesy of USGS

When the ice dam broke and the lake burst through, water rushed into the narrow river valley in a great deluge that probably lasted several days. At its height, water might have been 1,500 feet high and raced along the Chuja River Valley at 90 miles per hour. In the valley and nearby regions, massive flooding formed giant bars of gravel and oddly ripped terrain. The Chuja Valley flood was possibly the largest of the numerous ice age inundations around the world.

A gigantic ice dam on the border between Idaho and Montana held back a huge body of water called Lake Missoula that was hundreds of miles wide and up to 2,000 feet deep. Around 13,000 years ago, the dam suddenly burst, sending glacial meltwaters gushing toward the Pacific Ocean. Along the way, the flood waters carved out one of the strangest landscapes the planet has to offer known as the Channeled Scablands (Fig. 9–14), whose distinctive suite of landforms are associated with the immense stream power expended by the flooding.

The British Isles had been settled by Stone Age peoples during the last ice age when the English Channel and the North Sea Basin were exposed above sea level as the ocean fell by as much as 400 feet. The English Channel itself might have been carved by late glacial catastrophic flood erosion during the collapse of the great ice sheets during the Pleistocene.

Iceland is a broad volcanic plateau of the Mid-Atlantic Ridge, which rose above the sea about 16 million years ago. The island is unique because it straddles a spreading ridge system, where the two plates of the Atlantic

Figure 9–15 Houses partially buried by tephra (volcanic fragments) in the eastern part of Vestmannaeyjar from the July 1973 eruption of Heimaey, Iceland. Courtesy of USGS

Basin and adjacent continents pull apart. A steep-sided, V-shaped valley runs northward across the entire length of the island and is one of the few expressions of a midocean rift on land. Many volcanoes flank the rift, making Iceland one of the most volcanically active places on Earth.

Icelandic volcanism produces glacier-covered volcanic peaks up to a mile in elevation and a high degree of geothermal activity. Although Iceland is fortunate to have such an abundant supply of energy, used for geothermal electrical generation and heating, it is not without its dangerous side effects, and the island has been plagued with frequent volcanic eruptions. The most destructive eruption in modern times destroyed a third of the fishing village of Vestmannaeyjar on the island of Heimaey, in 1973 (Fig. 9–15). Since then, at least a dozen or more major volcanic eruptions have stricken the tiny country.

In 1918, an eruption under a glacier unleashed a flood of meltwater called a glacier burst. In a matter of days, it released up to 20 times more water than the flow of the Amazon, the world's largest river. At least 13 similar underglacier eruptions have occurred during the last half of this century. A glacier burst is a sudden release of a great quantity of meltwater from a glacier or subglacial lake. Water accumulates in depressions within the ice margins, and at a critical stage it erupts through the ice barrier, sometimes producing a catastrophic flood. The process of accumulation and release

Figure 9–16 An ice cave on the Ross Ice Shelf, Antarctica. Photo by W. J. Collins, courtesy of U.S. Navy

might occur at almost regular intervals. The phenomenon is most common in Iceland, where it is associated with volcanic or fumarolic (volcanic steam) activity.

Water gushing from an underglacier eruption carves out an enormous ice cave. Geothermal heat beneath the ice creates a large reservoir of meltwater as much as 1,000 feet deep. A ridge of rock acts as a dam to hold back the water. When the dam suddenly breaks open, the flow of water forms a channel under the ice upward of 30 miles or more long.

Ordinary meltwater flowing from a glacier also carves ice caves (Fig. 9–16). Ice caves formed by glacial outwash streams can be followed upstream for long distances. Often the stream can be heard gurgling beneath the glacier from crevasses slicing through the ice. The swift-flowing stream carries a heavy load of sediment that is deposited at the mouth of the glacier.

Lahars are mudflows produced by volcanic eruptions. They can be initiated by a pyroclastic or lava flow moving across a glacier, rapidly melting it and releasing enormous quantities of sediment-laden water. The speed of a lahar depends mostly on the slope of the terrain and its fluidity, which is controlled by the water content. Lahars travel swiftly down valley floors for a distance of up to 50 miles or more at speeds exceeding 20 miles per hour. Lava flows extending onto glacial ice or snowfields produce floods as well as lahars, with flood-hazard zones often extending long distances down the valley below.

UNDERSEA STRUCTURES

During the height of the last ice age, about 10 million cubic miles of water were locked up in the continental ice sheets, which covered about a third of the landscape with an ice volume three times greater than at present. The massive continental glaciers sprawling over much of the Northern Hemisphere held enough water to lower the ocean as much as 400 feet, advancing the shoreline hundreds of miles seaward. When the glaciers melted, the ocean returned to near its present level, inundating what was once dry land.

A step on the continental shelf off the eastern United States can be traced for nearly 200 miles. It appears to represent the former ice age coastline, now deeply submerged. The shoreline of the eastern seaboard of the United States extended about halfway to the edge of the continental shelf, which stretches more than 600 miles eastward of the present coast. The drop in sea level exposed land bridges and linked continents, spurring the migration of plants and animals to other parts of the world.

Blue holes off the Bahamas formed during the height of the last ice age around 18,000 years ago, when the seas dropped several hundred feet, exposing the area well above sea level. Acidic rainwater seeping into the soil dissolved the limestone bedrock, creating vast subterranean caverns. Under the weight of the surface rocks, the roofs of the caverns collapsed,

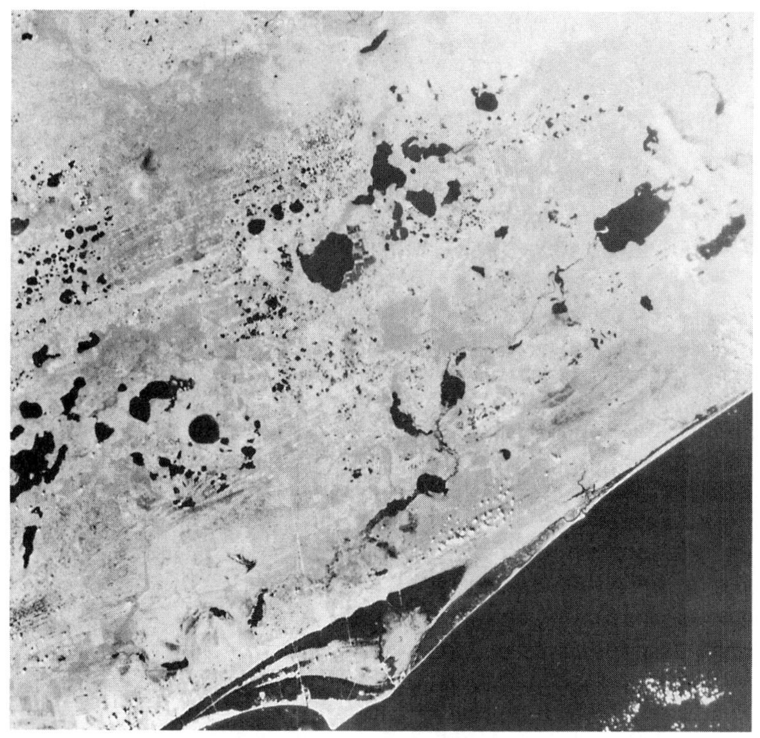

Figure 9–17 The Cape Kennedy area, Florida, showing numerous sinkholes filled with water. Courtesy of NASA

exposing huge, gaping pits called sinkholes (Fig. 9–17). The sinkholes filled with seawater when the great glaciers melted and the seas rose again.

When the ocean regressed as the sea level lowered dramatically during the last ice age, rivers flowing across the exposed seabed gouged out deep canyons in what was once the ocean floor. When the ocean transgressed upon formally exposed areas at the end of the ice age, it submerged the canyons. Submarine canyons carved into bedrock 200 feet below sea level can be traced to rivers on land, and many submarine canyons have heads near the mouths of large rivers.

Numerous submarine canyons cut through the continental margin and seafloor off eastern North America. Submarine canyons on continental shelves and slopes have identical features as river canyons, and some even rival the largest chasms on the continents. They have high, steep walls and an irregular floor that slopes continually outward. The canyons range upward of 30 miles and more in length, with an average wall height of about 3,000 feet. The Great Bahamas Canyon is one of the largest submarine canyons, with a wall height of 14,000 feet, making it over twice as deep as Arizona's Grand Canyon.

Several canyons slice through the continental shelf beneath the Bering Sea between Alaska and Siberia. About 75 million years ago, continental movements created the broad Bering Shelf rising 8,500 feet above the deep ocean floor. During the ice ages, the shelf was exposed as dry land when sea levels dropped several hundred feet. Terrestrial canyons cut deep into the shelf. When the ocean refilled again at the end of the last ice age, massive slides and mudflows swept down steep slopes on the shelf's edge, gouging out 1,400 cubic miles of sediment and rock.

The Arctic seafloor is crossed by grooves several miles long and up to 65 feet deep, carved by ancient, giant icebergs much larger than those of today. Evidence of megabergs challenges previous notions about the kind of ice that covered the Arctic region during the last ice age. It has long been thought that the Arctic held only thin sea ice at this time. But the discovery of parallel iceberg-carved channels on the ocean floor called plow marks raises the possibility that a floating ice sheet up to 2,000 feet or more thick covered much of the Arctic Ocean and connected the great glacial sheets over North America and northern Europe.

The iceberg plow marks were detected on the seabed northwest of Spitsbergen Island off Norway using side-scan sonar, which maps the ocean floor with sound waves. The plow marks occurred along the flanks of a submerged feature called the Yermak Plateau. Because the top of the plateau shows few gouges, it was probably beveled smooth by a thick ice shelf during the last ice age.

The Yermak Plateau iceberg plow marks are the first ones mapped in the Arctic Ocean and are perhaps the deepest iceberg drafts found anywhere. The depth of the scars indicates that some icebergs extended as much as 2,500 feet below the ocean surface. Only the present Antarctic produces icebergs approaching this size, with the deepest keels possibly reaching 1,300 feet below the waterline.

CORAL REEFS

The lack of accurate dates for the ice ages made testing glacial theories difficult. However, indirect evidence was secured by dating sea level fluctuations as indicated by coral terraces in the tropics. The corals constructed barrier reefs and atolls, which are massive structures composed of calcium carbonate lithified into limestone. The most remarkable feature of coral colonies is their ability to build huge calcareous skeletons, weighing several hundred tons. The major structural feature of a living reef is a coral rampart that reaches almost to the surface of the sea. It consists of large rounded coral heads and a variety of branching corals (Fig. 9–18).

The Great Barrier Reef (Fig. 9–19) along the northeast coast of Australia forms an underwater embankment more than 1,200 miles long, up to 90 miles wide, and rises as much as 400 feet above the ocean floor. It is the

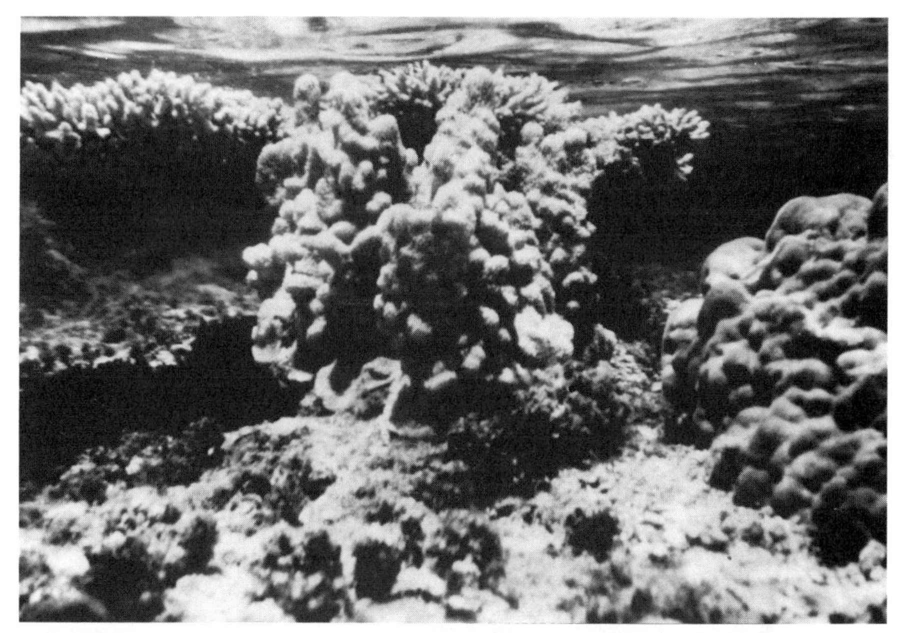

Figure 9–18 Coral at Bikini Atoll, Marshal Islands. Photo by K. O. Emery, courtesy of USGS

largest feature built by living organisms and one of the great wonders of the world. It is also a relatively young structure, formed largely during the Pleistocene ice ages, when seas fluctuated with the growth of continental glaciers over the last 3 million years.

The rise and fall of sea levels during the last few million years has produced coral terraces that resemble a stairstep running up an island or a continent. The drowned coral represents periods of extensive glaciation that dramatically dropped the level of the sea. The coral records also show when sea levels returned to normal, such as during the end of the ice age before last, when the ocean rose as high as it presently is by 130,000 years ago, marking the beginning of the Eemian interglacial. In Jamaica, almost 30 feet of reef have built up since the present sea level stabilized some 5,000 years ago, after the last of the glaciers retreated to the poles.

At the height of the ice age, approximately 5 percent of the planet's water was contained in glacial ice, lowering the sea level and expanding the land area up to 8 percent. Therefore, the shallow-water shelf areas shrunk about the same amount, as shorelines crept seaward. Coral, which lives only in warm surface waters, fluctuates in height in response to changing sea levels. The lowering of the sea by the buildup of glacial ice resulted in the erosion of coral reefs down to the new sea level.

The rising of the sea when the glaciers melted resulted in the growth of new coral on top of the old. Thus, the waxing and waning of the ice ages

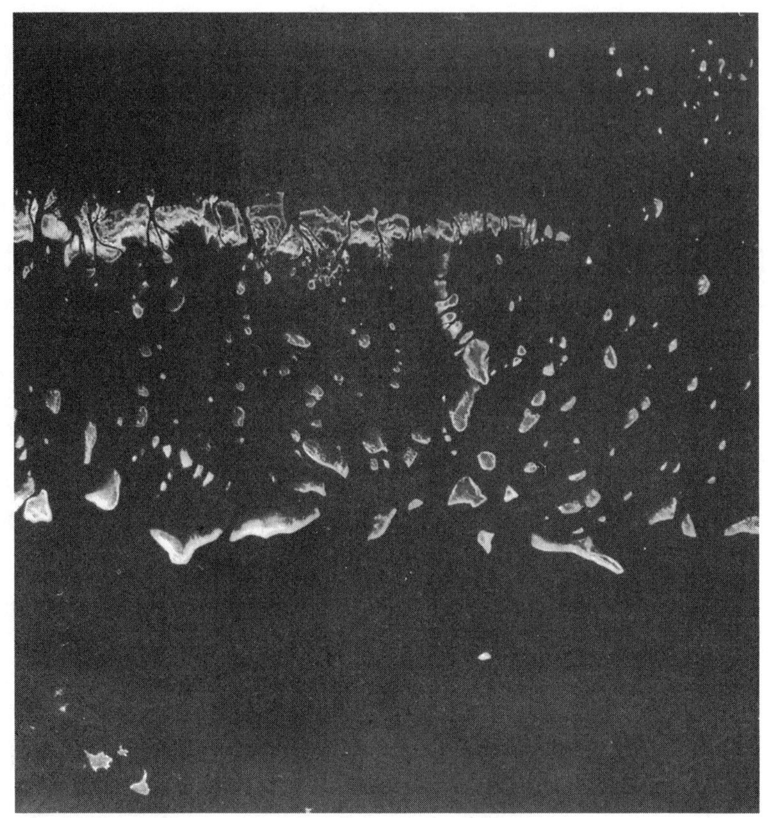

Figure 9–19 The Great Barrier Reef, northeast of Australia, is the world's most extensive reef system. Courtesy of NASA

formed a terrace of coral growth. When the corals were dated using radiometric dating techniques they provided a reliable chronology for the ice ages. A comparison between oxygen isotopes in the fossil coral also yielded the mean global temperature. While building their calcium carbonate shells, corals occasionally incorporate strontium, which resembles calcium in the ocean and occurs more frequently in colder water. These methods gave temperatures that were generally 5 degrees Celsius lower than at present.

Corals along with other sources are also good paleoclimate indicators. Like tree rings used to determine past climates, wide growth bands and dense skeletal colonies of corals mark times when the temperature, sea level, and climate were to their liking. By counting growth rings of ancient coral fossils an estimate of the number of days in a year is derived dating back to the beginning of the Cambrian period, 570 million years ago. At that time, the year had about 428 days, indicating the Earth was spinning much faster than it does today, which might explain in part why it was so much warmer.

10

GLACIAL DEPOSITS

Modern reminders of past ice ages come in a variety of forms. Most evidence for extensive glaciation is found in deposits of glacial rocks called erratics, dropstones, tillites, and moraines. Thick deposits of glacial sediment that buried older rocks, forming elongated hillocks aligned in the same direction, are called drumlins. Long, sinuous sand deposits called eskers are formed by glacial debris from outwash streams.

Glacial varves, alternating layers of silt and sand in ancient lake bed deposits, were laid down annually in lakes positioned below the outlets of glaciers. Most windblown sand deposits in the central United States, called loess, were laid down during the Pleistocene ice ages. In parts of the Arctic tundra, soil and rocks create strikingly beautiful and orderly patterns called frost polygons.

ERRATIC BOULDERS

In many regions once covered by glaciers, huge blocks of granite weighing several thousand tons, called erratic glacial boulders, are strewn across the mountainous areas (Fig. 10–1). Their existence helped lead to the recognition of widespread continental glaciation. In northern Europe, huge boul-

Figure 10–1 A perched erratic boulder left by the ice of the El Portal glaciation near the head of Little Cottonwood Creek, east of Army Pass, Inyo County, California. Photo by F. E. Matthes, courtesy of USGS

ders laid scattered about as though simply dumped there. Similar boulders were sprinkled across the slopes of the Jura Mountains in Switzerland. Geologists traced the boulders back to the Swiss Alps over 50 miles away.

Generally though, most geologists of the late 18th century thought the boulders were swept into the northlands by the Great Flood. However, granite erratic boulders resting on top of the limestones of the Jura Mountains in Switzerland were shown to be carried into the region by immense glaciers, lending persuasive evidence for the ice age.

The boulders were originally called drift because they appeared to have "drifted" in by water or on floating ice. Today, the term "glacial drift" (Fig. 10–2) refers to all rock material deposited by glaciers or glacier-fed streams and lakes, where the greatest thicknesses occur in buried valleys. Drift is divided into two types of material. One is till, deposited directly by glacial ice and shows little or no sorting or stratification (layering) as though haphazardly dumped. The other is stratified drift, which is a well sorted and layered material that has been moved and deposited by glacial meltwater. As the ice melts, meltwater streams rework part of the glacial material, some of which is carried into standing bodies of water.

Erratics are glacially transported boulders that might be embedded in glacial till or exposed on the ground surface. They range in size from pebbles to massive boulders and can travel as far as 500 miles or more.

Erratics composed of distinctive rock types can be traced to their place of origin and serve as indicators of glacial flow direction.

An indicator boulder is a glacial erratic of known origin used to locate the source area and transport distance for any given glacial till. Its identifying features include a distinctive appearance, unique mineral assemblage, or characteristic fossil continent. For example, erratics containing native copper torn from an outcrop in northern Michigan were scattered from Iowa to Ohio.

Indicator boulders are often arranged in a boulder train, which is a line or series of rocks derived from the same bedrock source and extends in the direction of glacial movement. A boulder fan is a fan-shaped area containing distinctive erratics derived from an outcrop at the apex of the fan. The angle at which the margins diverge measures the maximum change in the direction of glacial motion.

At the north end of Death Valley, California, lies a lake bed called the Racetrack (Fig. 10–3), which turns into a shallow lake a few times a year. The area is famous for its mysterious moving boulders that leave tracks a

Figure 10–2 An exposure of drift in a bank of Evans Creek deposited at or near the terminus of a valley glacier in the Carbon River Valley, Pierce County, Washington. Photo by D. R. Crandell, courtesy of USGS

Figure 10–3 The Racetrack showing mud-cracked playa and pebble-sized playa scrapers that dug furrows, Death Valley National Monument, Inyo County, California. Photo by J. F. McCallister, courtesy of USGS

fraction of an inch deep. High winds whistling off the nearby mountains appeared to push the boulders through the muddy lake bed after a soaking rain. But the largest boulder that has moved weighs some 700 pounds, much too heavy to be pushed around by the wind.

However, the wind might be able to move the boulders with a little help from ice. If a thin layer of ice formed on the lake bed after a winter rain, it would lift the boulders slightly, reducing the contact between the rocks and the mud. With the aid of the wind, the ice sheet would scoot across the lake bed with embedded boulders etching patterns in the mud.

DROPSTONES

Certain erratics found in nonglacial-transported marine sediments might have been deposited by icebergs. Large, out-of-place boulders strewn across the central desert of Australia suggest that ice existed there even during the warm Cretaceous period, about 100 million years ago. At this time, Australia was still attached to Antarctica and straddled the Antarctic Circle (Fig. 10–4), inside of which winters were sunless and cold. Maintaining wintertime temperatures above freezing in the interiors of large continents at high latitudes is difficult because they do not receive warmth from the ocean.

Like most continents of the Cretaceous, the interior of Australia contained a large inland sea. The continents were generally flatter and sea levels were hundreds of feet higher than today. Sediments settling on the

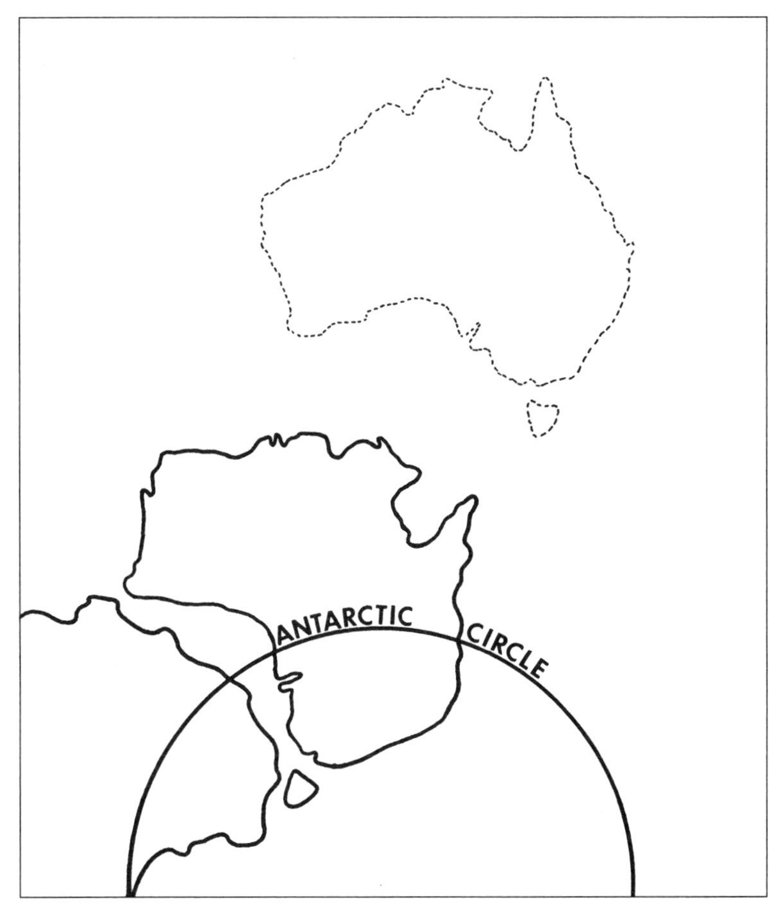

Figure 10–4 Location of Australia still attached to Antarctica during the middle Cretaceous with respect to the Antarctic Circle. Dashed lines are present location.

floor of the basin lithified into sandstone and shale and were later exposed when the land uplifted and the sea departed at the end of the period.

After Australia had drifted into the subtropics, the central portion of the continent became a large desert. Lying in the middle of the sedimentary deposits are curious-looking boulders of exotic rock called dropstones that measured as much as 10 feet across and came from a great distance away. Rivers or mudslides could not have carried the boulders into the middle of the basin because such torrents would have disturbed the smooth sediments composed of fine-grained sandstone and shale.

The appearance of these strange boulders out in the middle of nowhere suggests they rafted out to sea on slabs of drift ice. When the ice melted, the huge rocks simply dropped to the ocean floor, where their impacts disturbed the underlying sediment layers. The boulders apparently were not dropped by permanent glaciers but by seasonal ice packs that formed in winter. During the cold winters, portions of the interior coastline froze into pack ice. Rivers of broken ice then flowed into the inland sea, carrying with them embedded boulders dropped more than 60 miles from shore.

Evidence of ice-rafting of boulders during the Cretaceous also exists in glacial soils in other areas of the world such as the Canadian Arctic and Siberia. This suggests that the high latitudes still had cold climates, in which ice formed easily even during one of the warmest periods in Earth's history. Boulders were also found in sediments from other warm periods as well. The same ice-rafting process is occurring even today in the Hudson Bay.

In late 1993, off the southeast coast of Greenland, an international drilling expedition of the Ocean Drilling Program recovered layers of marine sediments that contained large rocks among fine-grained, muddy sediment. The discovery suggested that ice covered sections of Greenland much earlier than previously suspected. The new evidence hints that some ice sheets in the Northern Hemisphere began growing as early as 7 million years ago.

The large rocks in the drill cores provided clues that Greenland had at least a partial covering of ice during this time. The rocks indicated that icebergs plied this part of the North Atlantic, giving boulders a ride within the floating ice. When the icebergs calved off Greenland and drifted out to sea, they dropped boulders enroute as they melted.

Icebergs invading the North Atlantic during the last ice age spanned the entire region from Labrador to Europe. Before breaking off the continental ice sheet, the icebergs scraped up rocky debris from the land beneath the glacier and deposited it as dropstones when the ice melted. At least six times during the 100,000-year ice age, great armadas of icebergs calved off the ice sheet in Canada. During warmer intervals of about 5,000 to 10,000 years apart, icebergs suddenly invaded the ocean in vast numbers, dropping trails of boulders along the way.

TILLITES AND MORAINES

Most evidence for widespread glaciation exists in deposits of glacial rocks called tillites and moraines. The existence of ancient tillites and moraines found on every continent suggests that five or more major periods of glaciation have occurred over the last 2.4 billion years. Indeed, contemporaneous glacial deposits on all the southern continents indicate they had once moved en masse over the South Pole, which provided the clinching evidence for the existence of the great southern continent Gondwana.

Glacial till, also called boulder clay, is nonstratified material deposited directly by glacial ice. It comprises clay and intermediate-sized boulders that are usually angular because they have undergone little or no river transportation, which causes abrasion. Basal till carried in the base of the glacier was usually laid down under it. Ablation till, carried on or near the surface of the glacier, was deposited when it melted. The surface cover of sediment also might have protected the glacier from the heating rays of the sun. Some sun-heated rocks might have sunk into the glacier, forming deep depressions.

Much of the upper midwestern and northeastern parts of the United States were overrun by thick glaciers during the last ice age, and glacially derived sediments covered much of the landscape, burying older rocks under thick layers of till. Tillites are sedimentary rocks formed by the compaction and cementation of glacial tills. They are a mixture of boulders, pebbles, and clay deposited by glacial ice and consolidated into solid rock.

Thick sequences of Precambrian tillites are found on every continent (Fig. 10–5). In the Lake Superior region of North America, tillites up to 600 feet thick range from east to west for 1,000 miles or more. In northern Utah, tillites form impressive deposits as much as 12,000 feet thick. The various layers of sediment provide evidence for a series of ice ages following one after another in quick succession. Similar tillites exist among Precambrian rocks in Norway, Greenland, China, India, southwestern Africa, and Australia. In Australia, Permian marine sediments were interbedded with glacial deposits, and tillites were separated by seams of coal, indicating that periods of glaciation were interspersed with warm interglacial spells, when extensive forests grew.

Figure 10–5 Precambrian tillite of the Chocolay Group, Michigan.
Photo by W. F. Cannon, courtesy of USGS

TABLE 10–1 CONTINENTAL DRIFT

Geologic division (millions of years)		Gondwana	Laurasia
Quaternary	3		Opening of Gulf of California
Pliocene	11	Begin spreading near Galapagos Islands	Change spreading directions in eastern Pacific
		Opening of the Gulf of Aden	
Miocene	26		Birth of Iceland
		Opening of Red Sea	
Oligocene	37		
		Collison of India with Eurasia	Begin spreading in Arctic Basin
Eocene	54		Separation of Greenland from Norway
Paleocene	65	Separation of Australia from Antarctica	
			Opening of the Labrador Sea
		Separation of New Zealand from Antarctica	
			Opening of the Bay of Biscay
		Separation of Africa from Madagascar and South America	Major rifting of North America from Eurasia
Cretaceous	135		
		Separation of Africa from India, Australia, New Zealand, and Antarctica	
Jurassic	180		
			Begin separation of North America from Africa
Triassic	250		

In South Africa, the Karroo Series, a sequence of late Paleozoic tillites interbedded with coal beds, extend over an area of several thousand miles, reaching a total thickness of 20,000 feet. Between layers of coal, which is among the best in Africa, are fossil plant leaves of the extinct fern glossopteris, which only lived on the southern continents and is among the best evidence for the theory of continental drift.

Of all glacial landforms, perhaps the simplest are moraines. They are accumulations of rock material carried by a glacier and deposited in a regular, usually linear pattern that makes a recognizable landform. The sediment ranges in size from sand to boulders and shows no sorting or bedding, which requires the influence of water. The rocks are generally faceted (planed off) and striated by abrasion during transport.

The different types of moraines are named according to their position in relation to the glacier. A ground moraine is an irregular carpet of till deposited under a glacier and mainly composed of clay, silt, and sand. It is the most prevalent type of continental glacial deposit. A glacier overloaded at the base might drop some of its rock along the floor and move over it. When retreating, the glacier leaves additional rock waste, usually encased in sandy clay called boulder clay. Blocks of ice buried in the ground moraine leave depressions that might become small ponds after the ice melts.

Terminal moraines (Fig. 10–6) are ridges of erosional debris deposited by the melting forward margin of a glacier that has paused long enough for till to accumulate. A string of terminal moraines creates a broken line of irregular hills stretching along the former edge of the North American ice sheet from Cape Cod to the Rocky Mountains. A terminal moraine is a

Figure 10–6 Terminal moraine of a recent glacier on the east side of Mount Blanca, Costilla County, Colorado. Photo by W. T. Lee, courtesy of USGS

Figure 10–7 Mount Athabaska showing lateral moraine along the right side of Athabaska Glacier and directly in front of the mountain, Alberta province, Canada. Photo by F. O. Jones, courtesy of USGS

ridgelike mass of glacial debris formed by the foremost glacial snout and deposited at the outermost edge of glacial advance. Because the snout of a glacier is curved, the terminal moraine also curves down valley and might form lateral moraines up the side (Fig. 10–7).

Parallel lobes of moraine indicate a fluctuation in the position of the ice. A recessional moraine is a secondary terminal moraine deposited during a temporary halt in the retreat of a glacier. Thus, a series of these moraines show the history of glacial retreat. The moraines look as if they were bulldozed into place. However, glaciers are poor at pushing obstacles forward and are better at running over them. They act more like conveyor belts picking up debris beneath them and transporting the embedded sediment hundreds of miles, dropping it out at the front edge of the glacier when the ice melted.

The process is continuing even today. For over a hundred years, outflow glaciers have been steadily retreating, and terminal moraines mark their farthest advance. The various limits of retreat can be identified by studying the growth of lichens on rocks scattered about at the base of a glacier. This type of study is known as the science of lichenometry.

Figure 10–8 Folded morainic bands on Malaspina Glacier with Mount St. Elias on the skyline, Alaska Gulf region. Photo by D. J. Miller, courtesy of USGS

A lateral moraine is debris derived from erosion and avalanches from the valley wall onto the edge of the glacier. It forms a long ridge when the glacier recedes. The sides of a valley glacier are outlined by the ridges of rock debris torn from the valley walls and fallen from the cliffs above. When two valley glaciers merge into a single ice stream, the adjacent lateral moraines join in a central dark streak called a medial moraine (Fig. 10–8). It is an enlarged zone of debris formed when lateral moraines join at the intersection of two glaciers. A medial moraine forms a ridge running approximately parallel to the direction of ice movement.

DRUMLINS AND ESKERS

In many areas of the high northern latitudes, glacial ice stripped off entire layers of sediment, leaving behind bare bedrock. In other areas, thick deposits of glacial till created elongated hillocks aligned in the same direction called drumlins (Fig. 10–9). Drumlins rarely stand alone but accumulate in fields often containing thousands of glacier-made hills. The drumlin fields generally occur within a narrow band set back a short distance from a glacier's terminal moraine. The drumlins formed under the margins of glaciers or by the erosion of older moraines after reglaciation. Most drumlins are composed of clayey till, some have bedrock cores, and others contain sand and gravel sediments.

The drumlin's long axes are approximately parallel to the direction of ice movement. Drumlins are typically steepest and highest at the end facing the advancing ice and gently slope in the direction of the movement of a huge lobe of the continental ice sheet, which splayed outward as it plowed south. As a result, drumlins are tall and narrow at the upstream end of the glacier and slope to a low, broad tail at the downstream end. The hills appear in concentrated fields in North America, Scandinavia, Britain, and other areas once covered by glacial ice. Drumlin fields might contain as many as 10,000 knolls, looking much like rows of eggs lying on their sides.

Despite their distinctive appearance, drumlins are probably the least understood of all glacial landforms. They seem to have formed when the ice sheets contorted or deformed wet sediments lying beneath their bases. The sediments inside a drumlin often form complex, swirling layers, indicating they were stretched and sheared by moving ice. How they attained their characteristic oval shape still remains a mystery.

Similar to drumlins are roches moutonnées (Fig. 10–10) from the French words for "fleecy rock." The term was applied to glaciated outcrops because they resemble the backs of sheep in form and texture, thus prompting the name sheepback rock. It is a glaciated bedrock surface in the form of

Figure 10–9 A drumlin south of Newark, Wayne County, New York. Photo by G. K. Gilbert, courtesy of USGS

Figure 10–10 Polished and striated roches moutonnées developed on the Trommald Formation, Cross Wing County, Minnesota. Photo by R. G. Schmidt, courtesy of USGS

asymmetrical mounds of varying shapes. The up-glacier side has been glacially scoured and smoothly abraded. The down-glacier side has steeper, jagged slopes, resulting from glacial plucking, an erosional process by which glaciers dislodge and transport fragments of bedrock. The fragments might have been pried loose by the plastic flow of the ice around them and became part of the moving glacier. The ridges dividing the two sides of a roche moutonnée are perpendicular to the general flow of the ice sheets. Such landscapes are characteristic of glaciated Precambrian shields that make up the cores of the continents.

Figure 10–11 An esker in Dodge County, Wisconsin. Photo by W. C. Alden, courtesy of USGS

Figure 10–12 Chamouni Valley, France-Switerzerland, showing the milky white water of the glacier streams. Photo by J. A. Holes, courtesy of USGS

Eskers (Fig. 10–11), from a Celtic word for "mountain ridge," are long, narrow, sinuous or straight ridges comprised of poorly sorted and stratified glacial meltwater deposits of sand and gravel. The long, winding sand deposits were formed by glacial debris carried out from beneath the ice by outwash streams. The meltwater flowing out of a glacier is usually milky white (Fig. 10–12) from suspended fine material called rock flour, produced by the abrasion of the glacial ice. The material resembles clay but is actually composed of fine mineral fragments that form glacial varves.

Eskers usually occur in areas once occupied by the ground moraine of a continental glacier. Most eskers appear to have been deposited in channels beneath or within slow-moving or stagnant glacial ice. Their general orientation runs at right angles to the glacial edge. At the margin of a glacial lake, they might take the form of river deltas. Some eskers originate on the ice and contain ice cores.

Most eskers are in the shape of winding, steep-walled ridges that extend noncontinuously up to 500 miles in length but seldom exceed more than 1,000 feet in width and 150 feet in height. They were probably created by streams flowing through tunnels beneath or within the ice sheet. When the ice melted, the old stream deposits were left standing as a ridge. Well-known esker areas exist in Maine, Canada, Sweden, and Ireland.

Figure 10–13 Kames on the north side of Happy Valley Creek, Greenland. Photo by R. B. Colton, courtesy of USGS

Kames (Fig. 10–13), from the Scottish word for "crooked and winding," are mounds composed chiefly of stratified sand and gravel formed at or near the snout of an ice sheet or deposited at the margin of a melting glacier. Like eskers, kames occur in areas where large quantities of coarse material are available because of slow melting of stagnant ice. Meltwater must be present in sufficiently large quantities to redistribute the debris and deposit the sediments at the margins of the decaying ice mass.

Many kames probably formed when streams cascaded off the top of a glacier onto the bare ground, dropping sediment in a pile. Some kames formed in moulins, which are holes bored straight down through the ice to the bottom. When meltwater from the top of the glacier plunges into the moulin, it deposits a load of sediment that piles up into a cone.

Most kames are low, irregularly conical mounds of roughly layered glacial sand and gravel that often occur in clusters. They associate with the terminal moraine region of both valley and continental glaciers and appear

Figure 10–14 Kame and kettle topography of continental ice moraine in the vicinity of Spider Lake, Glacier County, Montana. Photo by H. E. Malde, courtesy of USGS

to represent sediment fillings of openings in stagnant ice. Others might have formed where glacial streams escaped from the ice. Streams flowing between the sides of a glacier and the enclosing valley walls form kame terraces that stand above the valley floor after the ice has melted.

Streams issuing from a glacier are heavily loaded with sediment that is deposited rapidly in a complex braided pattern of channels, which spread out the debris in a series of alluvial fans that coalesce into a flat, broad outwash plain that might be pock-marked with kettles. The irregular surface of the terminal zone of a continental glacier produces a kame and kettle topography (Fig. 10–14) with alternate hills and hollows. It also might take the form of hummocky ground peppered with knolls or knobs.

GLACIAL VARVES

The swift-flowing stream of meltwater carries a heavy load of sediment deposited at the mouth of a glacier, forming glacial varves, which are alternating layers of sand and silt laid down annually in a lake below the outlet of a glacier. Varves generally develop on the floors of cold freshwater lakes fed by intermittently flowing glacial meltwater streams. They rarely occur in salty or brackish water. Each summer when the glacial ice melts, turbid meltwater discharges into the lake and sediments settle out differentially, with coarse ones landing on the bottom first, forming a banded deposit.

Glacial varves, also called rhythmites, are regularly banded deposits developed by cyclic sedimentation, with fine-grained, dark laminae alternating with coarse-grained lighter layers. Each individual pair of laminae, called a couplet, represents a one-year cycle of fast melting during summer and slow melting during winter. Therefore, they can be used for dating purposes, especially for the late Pleistocene.

The varying widths of the varves were thought to represent stages in the solar cycle when an increase in sunspot activity warmed the climate slightly higher, melting more ice than usual. In lake bed deposits north of Adelaide in South Australia, glacial varves, dating to the Precambrian ice age about 680 million years ago, show distinct bands of varying widths of mud and silt with lighter bands interspersed between darker ones.

Each summer when the glacial ice melted, meltwater loaded with sediment discharged into a lake below the glacier, and the sediments settled out to form a stratified deposit. When the lake bed sediments were being laid down, thick ice sheets covered much of Australia (Fig. 10–15) during possibly the most severe ice age the planet has ever known, when glaciers covered the land surface almost to the equator.

Increased solar activity caused a corresponding rise in the temperature on Earth, which entailed a greater annual discharge of glacial meltwaters and the deposition of thicker, darker layers of sediment. The varves ap-

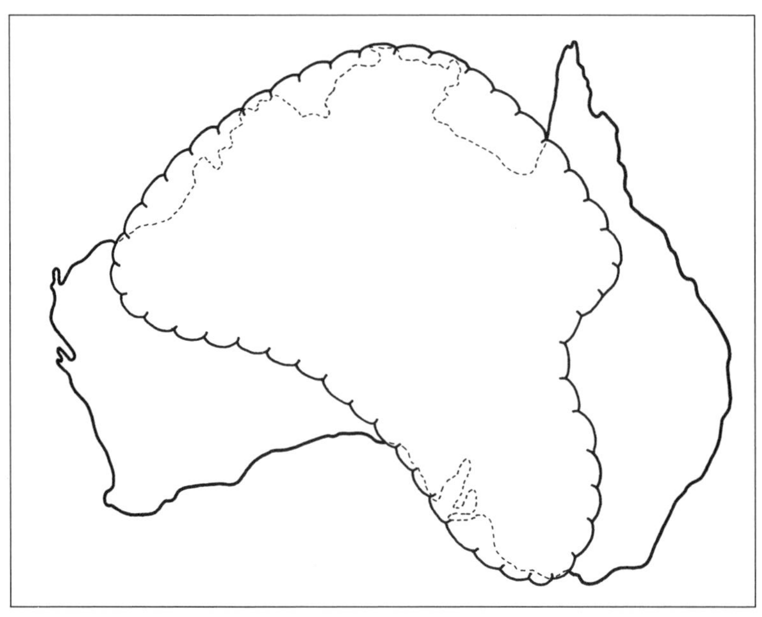

Figure 10–15 The late Precambrian ice sheet overlying Australia.

peared to mimic the solar cycle, with periods of roughly 11, 22, and 90 years. Also, periods of 145 and 290 years appeared to match modern rhythms in some tree ring climate records.

The correspondence between sunspot activity and varve thickness provided a strong argument for a link between solar activity and terrestrial climate 680 million years ago. However, newly discovered sediments with far thicker sedimentary cycles containing 14 to 15 laminations per cycle might better represent the lunar tidal cycle. Today the lunar cycle has a period of about 19 years, indicating that the moon was much closer to the Earth during the late Precambrian.

Banded lake bed sediments from the Newark Basin in northern New Jersey, date to about 200 million years ago. They are repetitive sequences corresponding to cycles of varying lake depth, which was influenced by the changing distribution of sunlight over the Earth during any given season. The cycles closely resemble the periods of precession of the Earth's axis and eccentricity of its orbit. Similar cycles occur in sediments throughout the world, suggesting that orbital variations might have operated on the climate throughout geologic history.

Another stratified deposit that could well be the most beautiful, economically important, and enigmatic rocks ever created on the planet are the banded iron formations. Precambrian mineral deposits are generally characterized as bedded or stratified. Iron, the fourth most abundant element in the Earth's crust, was leached from the continents and dissolved in seawater under reducing (lacking oxygen) conditions. When the iron reacted with oxygen in the ocean, it precipitated in vast deposits on shallow continental margins.

Alternating bands of iron-rich and iron-poor sediments gave the ore a banded appearance, thus prompting the name banded iron formation. These deposits, mined extensively throughout the world, provide over 90 percent of the minable iron reserves. In effect, biologic activity was responsible for the iron deposits, since photosynthetic organisms produced the oxygen. By the middle Precambrian, photosynthesis generated enough oxygen to react with iron on a grand scale.

The banded iron deposits formed some 2 billion years ago at the height of the first ice age. For unknown reasons, major episodes of iron deposition coincided with periods of glaciation. The average ocean temperature was probably warmer than today. When warm ocean currents rich in iron and silica flowed toward the glaciated polar regions, the suddenly cooled waters could no longer hold minerals in solution. Their precipitation formed alternating layers due to the difference in settling rates between silica and iron, the heavier of the two minerals.

LOESS DEPOSITS

During periods of glaciation, the climate was both colder and drier. The lower precipitation levels expanded the area of arid zones and increased the size of deserts. Strong winds blowing across desert regions produced gigantic sandstorms and raised huge clouds of dust. High amounts of dust

Figure 10–16 An exposure of loess standing in vertical cliffs, Warren County, Mississippi. Photo by E. W. Shaw, courtesy of USGS

suspended in the atmosphere blocked out sunlight and shaded the planet, keeping it cool. The fallout from the dust clouds also landed on the Greenland and Antarctic ice sheets, providing an excellent paleoclimate indicator.

Most windblown sediments accumulated into thick deposits of loess (Fig. 10–16), a fine-grained, loosely consolidated, sheetlike deposit, which on outcrop often shows thin, uniform bedding. Loess is comprised of angular particles of equal grain size composed of quartz, feldspar, hornblende, mica, and bits of clay. It is usually a buff to yellowish brown loamy deposit. Loess often contains the remains of grass roots, and like mud bricks deposits can stand in nearly vertical walls despite their weak cohesion.

Loess deposits lie downwind from glaciated areas. They appear to have formed when glacial retreat left large unvegetated areas that were subjected to wind erosion. Deposits are common in North America, Europe, and Asia, where China holds the largest loess deposits in the world. Most loess deposits in the central United States were laid down during the Pleistocene ice ages. They cover parts of the Mississippi Valley and portions of the Columbia Plateau.

During an ice age, regions not covered by glaciers dried out, causing widespread desertification. In dry regions where dust storms are prevalent, the wind transports large quantities of loose sediment. Windblown sediments landing in the ocean slowly build deposits of abyssal red clay, whose color signifies its terrestrial origin. The windblown sediment also contains significant amounts of iron, which in the ocean is an important nutrient that supports prolific blooms of marine plankton. The plankton draw down atmospheric carbon dioxide, and the loss of this important greenhouse gas

Figure 10–17 A terrace structure in the leeward side of sand dunes, Point Año Nuevo, San Mateo County, California. Photo by R. Arnold, courtesy of USGS

encourages colder climatic conditions, which in turn cause more drying and additional atmospheric dust, helping to sustain the ice age.

Along with loess are dune deposits, composed of desert sand, which when lithified show a distinct dune structure on outcrops. The sediment grains of desert deposits are often frosted due to the constant motion of the sand, which causes abrasion. The dunes move across the desert floor in response to the wind by a process known as saltation, by which sand grains in motion dislodge one another and become airborne for an instant. In this manner, sand dunes engulf everything in their paths as they march across the desert floor (Fig. 10–17).

FROST POLYGONS

Most of the ground in the Arctic tundra is permafrost and frozen year-round, and only the top few inches of soil thaws during the short summers. Although the ground is bathed in 24-hour sunlight, the soil temperature seldom rises above the freezing point of water. This is because as ice thaws, it absorbs heat from its environment.

As the ground begins to thaw in the Arctic summer, the retreating snows unveil a bizarre assortment of rocks arranged in a honeycomblike network,

Figure 10–18 Stone polygons east of Maclaren River, Clearwater Mountains, Valdez Creek District, Alaska. Photo by C. Wahrhaftig, courtesy of USGS

Figure 10–19 Large folds of the Malaspina Glacier, Alaska. Courtesy of USGS

giving the landscape the appearance of a tiled floor. Soil and rocks of the tundra are often fashioned into strikingly beautiful and orderly patterns that have confronted geologists for centuries. Even on Mars, spacecraft images have revealed furrowed rings, polygonal fractures, and ground-ice patterns of every description.

The polygons range in size from a few inches across when composed of small pebbles to several tens of feet wide when large boulders form protective rings around mounds of soil (Fig. 10–18). These patterns occur in most of the northern lands and alpine regions, where the soil is exposed to moisture and seasonal freezing and thawing cycles. The polygons probably resulted from similar processes that cause frost heaving, which thrusts boulders upward through the soil, a major annoyance to northern farmers. Rocks also penetrate highway pavement, and fence posts thrust completely out of the ground by frost heaving.

The boulders move through the soil by a pull from above and by a push from below. If the top of the rock freezes first, the expanding frozen soil pulls it upward. When the soil thaws, sediment gathers below the rock and it settles at a slightly higher level. The expanding frozen soil lying below also heaves the rock upward. After several frost-thaw cycles, the boulder finally comes to rest on the surface.

The regular polygonal patterned ground might have formed by the movement of soil of mixed composition upward toward the center of the mound and downward under the boulders, making the soil move in convective cells. The coarser material composed of gravel and boulders gradually shoves radially outward from the central area, leaving the finer materials behind. The arrangement of rocks in this manner suggests that the soil is being churned up by convection.

Other geometric designs in the Arctic soil include steps, stripes, and nets, which lie between the circles and polygons. These forms can reach 150 feet in diameter. Relics of ancient surface patterns, measuring up to 500 feet, have been found in former permafrost regions. These among many other mysterious features make the Arctic one of the most fascinating places in the world (Fig. 10–19).

GLOSSARY

ablation	erosion of a glacier by such processes as melting and evaporation
abrasion	erosion by friction, generally caused by rock particles carried by running water, ice, and wind
abyss	the deep ocean, generally over a mile in depth
aerosol	a mass of solid or liquid particles dispersed in air
age	a geological time interval that is smaller than an epoch such as an ice age
albedo	the amount of sunlight reflected from an object and dependent on its color and texture
alluvium	stream-deposited sediment
alpine glacier	a mountain glacier or a glacier in a mountain valley
aphelion	the point at which the orbit of a planet is at its farthest distance from the sun. In the Earth's case, it occurs in early July
arête	a sharp-crested ridge formed by abutting cirques
ash fall	the fallout of small, solid particles from a volcanic eruption cloud

asthenosphere	a layer of the upper mantle, roughly between 50 and 200 miles below the surface, that is more plastic than the rock above and below and might be in convective motion
atmospheric pressure	the weight per unit area of the total mass of air above a given point; also called barometric pressure
aurora	luminous bands of colored light seen near the poles due to cosmic-ray bombardment of the upper atmosphere
axis	a straight line about which a body rotates
barrier island	a low, elongated coastal island that parallels the shoreline and protects the beach from storms
basalt	a dark volcanic rock that is usually quite fluid in the molten state
basement rock	subterranean igneous, metamorphic, granitized, or highly deformed rock underlying younger sediments
bedrock	solid layers of rock lying beneath younger material
bicarbonate	an ion created by the action of carbonic acid on surface rocks. Marine organisms use the bicarbonate along with calcium to build supporting structures composed of calcium carbonate.
biogenic	sediments composed of the remains of plant and animal life such as shells
biomass	the total mass of living organisms within a specific habitat
biosphere	the living portion of the Earth that interacts with all other biological and geologic processes
calcite	a mineral composed of calcium carbonate
caldera	a large pitlike depression at the summits of some volcanoes and formed by great explosive activity and collapse
calving	formation of icebergs by breaking off of glaciers entering the ocean
Cambrian explosion	a rapid radiation of species that occurred as a result of a large adaptive space, including a large number of habitats and a mild climate

carbonaceous	a substance containing carbon, namely sedimentary rocks such as limestone and certain types of meteorites
carbonate	a mineral containing calcium carbonate such as limestone and dolostone
carbon cycle	the flow of carbon into the atmosphere and ocean, the conversion to carbonate rock, and the return to the atmosphere by volcanoes
catastrophism	a theory proposing that recurrent, violent global events cause sudden disappearance and appearance of species
Cenozoic	an era of geologic time comprising the last 65 million years
cirque	a glacial erosional feature, producing an amphitheater-like head of a glacial valley
coal	a fossil fuel deposit originating from metamorphosed plant material
col	a saddle-shaped mountain pass formed by two opposing cirques
comet	a celestial body believed to originate from a cloud of comets that surrounds the sun and develops a long tail of gas and dust particles when traveling near the inner Solar System
condensation	the process whereby a substance changes from the vapor phase to liquid or solid phase; the opposite of evaporation
conductivity	the quality of an object to transmit energy
continent	a landmass composed of light, granitic rock that rides on the denser rocks of the upper mantle
continental drift	the concept that the continents have been drifting across the surface of the Earth throughout geologic time
continental glacier	an ice sheet covering a portion of a continent
continental margin	the area between the shoreline and the abyss

continental shelf — the offshore area of a continent in shallow sea

continental shield — ancient crustal rocks upon which the continents grew

continental slope — the transition from the continental margin to the deep-sea basin

convection — a circular, vertical flow of a fluid medium by heating from below. As materials are heated, they become less dense and rise, cool down, become more dense and sink.

coral — any of a large group of shallow-water, bottom-dwelling marine invertebrates that are reef-building colonies common in warm waters

cordillera — a range of mountains that includes the Rockies, Cascades, and Sierra Nevada in North America and the Andes in South America

core — the central part of the Earth, consisting of a heavy iron-nickel alloy; also a cylindrical rock sample

Coriolis effect — the apparent force that deflects wind and ocean currents, causing them to curve in relation to the rotating Earth

correlation — the tracing of equivalent rock exposures over distance usually with the aid of fossils

cosmic rays — high-energy charged particles that enter the Earth's atmosphere from outer space

craton — the stable interior of a continent, usually composed of the oldest rocks

crevasse — a deep fissure in the crust or a glacier

crust — the outer layers of a planet's or a moon's rocks

crustal plate — a segment of the lithosphere involved in the interaction of other plates in tectonic activity

density — the amount of any quantity per unit volume

diapir — the buoyant rise of a molten rock through heavier rock

diatoms	microplants whose fossil shells form siliceous sediments called diatomaceous earth
dropstone	a boulder embedded in an iceberg and dropped to the seabed when it melted
drought	a period of abnormally dry weather
drumlin	a hill of glacial debris facing in the direction of glacial movement
dune	a ridge of windblown sediments usually in motion
earthquake	the sudden rupture of rocks along active faults in response to geological forces within the Earth
eccentricity	deviation from a circular to an elliptical orbit
ecliptic	the plane of the Earth's orbit around the sun
ecosystem	a community of organisms and their environment functioning as a complete, self-contained biological unit
environment	the complex physical and biological factors that act on an organism to determine its survival and evolution
eolian	a deposit of windblown sediment
eon	the longest unit of geologic time—roughly about a billion years or more in duration
epoch	a geologic time unit shorter than a period and longer than an age
equinox	either of the two points of intersection of the sun's path and the plane of the Earth's equator
era	a unit of geologic time below an eon consisting of several periods
erosion	the wearing away of surface materials by natural agents such as wind and water
erratic boulder	a glacially deposited boulder far from its source
esker	a curved ridge of glacially deposited material
evaporation	the transformation of a liquid into a gas

evaporite	the deposition of salt, anhydrite, and gypsum from evaporation in an enclosed basin of stranded seawater
evolution	the changing of physical and biological factors with time
extinction	the loss of large numbers of species over a short geologic time
extraterrestrial	pertaining to all phenomena outside the Earth
extrusive	an igneous volcanic rock ejected onto the surface of a planet or moon
fault	a break in crustal rocks caused by Earth movements
fauna	the animal life of a particular area or age
fissure	a large crack in the crust through which magma might escape from a volcano
flora	the plant life of a particular area or age
fluvial	pertaining to being deposited by a river
foraminifera	calcium carbonate-secreting organisms that live in the surface waters of the oceans
formation	a combination of rock units that can be traced over distance
fossil	any remains, impression, or trace in rock of a plant or animal of a previous geologic age
frost heaving	the lifting of rocks to the surface by the expansion of freezing water
frost polygons	polygonal patterns of rocks from repeated freezing
fumarole	a vent through which steam or other hot gases escape from underground such as a geyser
geothermal	the generation of hot water or steam by hot rocks in the Earth's interior
geyser	a spring that ejects intermittent jets of steam and hot water
glacier	a thick mass of moving ice occurring where winter snowfall exceeds summer melting

glacière an underground ice formation

glacier burst a flood caused by an underglacier volcanic eruption

glossopteris a late Paleozoic plant that existed on the southern continents, but not found on the northern continents, thereby confirming the existence of Gondwana

Gondwana a southern supercontinent of Paleozoic time, comprised of Africa, South America, India, Australia, and Antarctica. It broke up into the present continents during the Mesozoic era.

graben an elongated, down-dropped block of crust bounded by faults

granite a coarse-grained, silica-rich rock, consisting primarily of quartz and feldspars. It is the principal constituent of the continents and believed to be derived from a molten state beneath the Earth's surface.

greenhouse effect the trapping of heat in the lower atmosphere principally by water vapor and carbon dioxide

hanging valley a glaciated valley above the main glaciated valley often forming a waterfall

heat budget the flow of solar energy through the biosphere

Holocene epoch a geological time covering the last 10,000 years commensurate with civilization

horn a peak on a mountain formed by glacial erosion

horst an elongated, uplifted block of crust bounded by faults

hot spot a volcanic center with no relation to a plate boundary; an anomalous magma generation site in the mantle

hyaloclastic basalt lava erupted beneath a glacier

hydrocarbon a molecule consisting of carbon chains with attached hydrogen atoms

hydrologic cycle the flow of water from the ocean to the land and back to the sea

hydrothermal relating to the movement of hot water through the crust

Iapetus Sea	a former sea that occupied a similar area as the present Atlantic Ocean prior to the assemblage of Pangaea
ice age	a period of time when large areas of the Earth were covered by massive glaciers
iceberg	a portion of a glacier calved off upon entering the sea
ice cap	a polar cover of snow and ice
igneous rocks	all rocks solidified from a molten state
infrared	heat radiation with a wavelength between red light and radio waves
insolation	all solar radiation impinging on a planet
interglacial	a warming period between glacial periods
intrusive	any igneous body that has solidified in place below the Earth's surface
iridium	a rare isotope of platinum, relatively abundant on meteorites
island arc	volcanoes landward of a subduction zone, parallel to a trench, and above the melting zone of a subducting plate
isostasy	a geologic principle that states that the Earth's crust is buoyant and rises and sinks depending on its density
isotope	a variety of an element with a different number of neutrons in the nucleus
jet stream	relatively strong winds concentrated within a narrow belt
kame	a steep-sided mound of moraine deposited at the margin of a melting glacier
karst	a terrain comprised of numerous sinkholes in limestone
kettle	a depression in the ground caused by a buried block of glacial ice
lateral moraine	the material deposited by a glacier along its sides
Laurasia	a northern supercontinent of Paleozoic time consisting of North America, Europe, and Asia

lava	molten magma that flows out onto the surface
lichenometry	the study of lichens to determine past positions of glaciers
limestone	a sedimentary rock consisting mostly of calcite from shells of marine invertebrates
lithosphere	the rocky outer layer of the mantle that includes the terrestrial and oceanic crusts. The lithosphere circulates between the Earth's surface and mantle by convection currents.
loess	a thick deposit of airborne dust
magma	a molten rock material generated within the Earth and is the constituent of igneous rocks
magnetic field reversal	a reversal of the north-south polarity of the magnetic poles
mantle	the part of a planet below the crust and above the core, composed of dense rocks that might be in convective flow
Maunder minimum	a period of unusually low sunspot activity from 1645 to 1715
mean temperature	the average of any series of temperatures observed over a period of time
megaherbivore	a large plant-eating animal such as an elephant or extinct mastodon
mesosphere	a region of the atmosphere between the stratosphere and thermosphere, extending 24 to 48 miles above the Earth's surface
Mesozoic	literally the period of middle life, referring to a period between 250 and 65 million years ago
metamorphism	recrystallization of previous igneous, metamorphic, or sedimentary rocks created under conditions of intense temperatures and pressures without melting

meteorite	a metallic or stony celestial body that enters the Earth's atmosphere and impacts on the surface
methane	a hydrocarbon gas liberated by the decomposition of organic matter and a major constituent of natural gas
Mid-Atlantic Ridge	the seafloor spreading ridge that marks the extensional edge of the North and South American plates to the west and the Eurasian and African plates to the east
midocean ridge	a submarine ridge along a divergent plate boundary where a new ocean floor is created by the upwelling of mantle material
moraine	a ridge of erosional debris deposited by the melting margin of a glacier
nova	a star that suddenly brightens during its final stages
Oort cloud	the collection of comets that surround the sun about a light-year away
ophiolite	masses of oceanic crust thrust onto the continents by plate collisions
orbit	the circular or elliptical path of one body around another
orogens	eroded roots of ancient mountain ranges
orogeny	a process of mountain building by tectonic activity
outgassing	the loss of gas from within a planet as opposed to degassing, the loss of gas from meteorites
ozone	a molecule consisting of three oxygen atoms in the upper atmosphere that filters out harmful ultraviolet radiation from the sun
paleomagnetism	the study of the Earth's magnetic field, including the position and polarity of the poles in the past
paleontology	the study of ancient life forms, based on the fossil record of plants and animals
Paleozoic	the period of ancient life, between 570 and 250 million years ago

Pangaea	an ancient supercontinent that included all the lands of the Earth
Panthalassa	a great world ocean that surrounded Pangaea
peridotite	the most common rock type in the mantle
perihelion	the point at which the orbit of a planet is at its nearest to the sun. In the Earth's case, it occurs in early January.
period	a division of geologic time longer than an epoch and included in an era
permafrost	permanently frozen ground in the Arctic regions
photosynthesis	the process by which plants create carbohydrates from carbon dioxide, water, and sunlight
phylum	a group of organisms that share similar body forms
phytoplankton	marine or freshwater microscopic single-celled, freely drifting plant life
pillow lava	lava extruded on the ocean floor giving rise to tabular shapes
plate tectonics	the theory that accounts for the major features of the Earth's surface in terms of the interaction of lithospheric plates
placer	a deposit of rocks left behind by a melting glacier; any ore deposit that is enriched by stream action
Pleistocene epoch	a geologic time commensurate with the ice ages, beginning about 3 million years ago
pluvial lake	lake formed by rainwater during an ice age
polar wandering	movement of the geographic poles
polynya	an open-water area in polar sea ice
pothole	deep depressions in the bedrock of a fast-flowing stream or beneath a waterfall
precession	the slow change in direction of the Earth's axis of rotation due to gravitational action of the moon on the Earth

precipitation	products of condensation that fall from clouds as rain, snow, hail, or drizzle; also the deposition of minerals from seawater
primordial	pertaining to the primitive conditions that existed during early stages of development
radiometric dating	the age determination of an object by chemical analysis of stable verses unstable radioactive elements
recessional moraine	a glacial moraine deposited by a retreating glacier
reef	the biological community that lives at the edge of an island or continent. The shells from dead organisms form a limestone deposit.
regression	a fall in sea level, exposing continental shelves to erosion
reversed magnetism	a geomagnetic field with a reverse polarity from that of the present one
revolution	the motion of a celestial body in its orbit as with the Earth around the sun
rift valley	the center of an extensional spreading center where continental or oceanic plate separation occurs
roche moutonnée	a knobby, glaciated, bedrock surface
rotation	the turning of a body about an axis
saltation	the movement of sand grains by wind or water
sandstone	a sedimentary rock consisting of cemented sand grains
seafloor spreading	a theory that the ocean floor is created by the separation of lithospheric plates along midocean ridges, with new oceanic crust formed from mantle material that rises from the mantle to fill the rift
seamount	a submarine volcano
sedimentary rock	a rock composed of fragments cemented together

shield — areas of exposed Precambrian nucleus of a continent

solstice — the occurrence twice yearly when the apparent distance of the sun from the equator is at its greatest

strata — layered rock formations; also called beds

stratosphere — the upper atmosphere above the troposphere, about 9 to 12 miles above sea level

striae — scratches on bedrock made by rocks embedded in a moving glacier

stromatolite — a calcareous structure built by successive layers of bacteria and have been in existence for the past 3.5 billion years

subduction zone — a region where an oceanic plate dives below a continental plate into the mantle. Ocean trenches are the surface expression of a subduction zone.

submarine canyon — a deep gorge residing undersea and formed by the underwater extension of rivers

sunspot — dark, cooler areas on the sun during intense solar activity

surge glacier — a continental glacier that heads toward the sea at a high rate of advance

tarn — a small lake formed in a cirque

tectonic activity — the formation of the Earth's crust by large-scale movements throughout geologic time

Tethys Sea — the hypothetical midlatitude region of the oceans separating the northern and southern continents of Laurasia and Gondwana

tillite — a sedimentary deposited composed of glacial till

transgression — a rise in sea level that causes flooding of the shallow edges of continental margins

tundra — permanently frozen ground at high latitudes and elevations

uniformitarianism	a theory that the slow processes that shape the Earth's surface have acted essentially unchanged throughout geologic time
varves	thinly laminated lake bed sediments deposited by glacial meltwater
volcanism	any type of volcanic activity, including volcanoes, fumaroles, and geysers

BIBLIOGRAPHY

DISCOVERING THE ICE AGES

Anderson, Don L. "The Earth as a Planet: Paradigms and Paradoxes." *Science* 223 (January 27, 1984): 347–354.

Badash, Lawrence. "The Age-of-the-Earth Debate." *Scientific American* 261 (August 1989): 90–96.

Cann, Joe and Cherry Walker. "Breaking New Ground on the Ocean Floor." *New Scientist* 139 (October 30, 1993): 24–29.

Covey, Curt. "The Earth's Orbit and the Ice Ages." *Scientific American* 250 (February 1984): 58–66.

Cullen, Christopher. "Was There a Maunder Minimum." *Nature* 283 (January 31, 1983): 427–428.

Culotta, Elizabeth. "Is the Geological Past a Key to the (Near) Future?" *Science* 259 (February 12, 1993): 906–908.

Fodor, R. V. "Explaining the Ice Ages." *Weatherwise* 35 (June 1982): 109–114.

Kerr, Richard A. "How High Was Ice Age Ice? A Rebounding Earth May Tell." *Science* 265 (July 8, 1994): 189.

——. "A Revisionist Timetable for the Ice Ages." *Science* 258 (October 1992): 220–221.

Matthews, Samuel W. "Ice on the World." *National Geographic* 171 (January 1987): 84–103.

Monastersky, Richard. "Ice Age Insights." *Science News* 134 (September 17, 1988): 184–186.

Historical Ice Ages

Allegre, Claude J. and Stephen H. Sneider. "The Evolution of the Earth." *Scientific American* 271 (October 1994): 66–75.

Allen, Joseph B. and Tom Waters. "The Great Northern Ice Sheet." *Earth* 4 (February 1995): 12.

Barron, Eric J. "Lessons from Past Climates." *Nature* 360 (December 10, 1992): 533.

Dalziel, Ian W. D. "Earth before Pangaea." *Scientific American* 272 (January 1995): 58–63.

Kasting, James F. "New Spin on Ancient Climate." *Nature* 364 (August 26, 1993): 759–761.

Kerr, Richard A. "The Whole World Had a Case of the Ice Age Shivers." *Science* 262 (December 24, 1993): 1972–1973.

Knoll, Andrew H. "End of the Proterozoic Eon." *Scientific American* 265 (October 1991): 64–73.

Larson, Roger L. "The Mid-Cretaceous Superplume Episode." *Scientific American* 272 (February 1995): 82–86.

Lemonick, Michael D. "The Ice Age Cometh?" *Time* 143 (January 31, 1994): 79–81.

Levinton, Jeffrey S. "The Big Bang of Animal Evolution." *Scientific American* 267 (November 1992): 84–91.

Waters, Tom. "Greetings from Pangaea." *Discover* 13 (February 1992): 38–43.

York, Derek. "The Earliest History of the Earth." *Scientific American* 268 (January 1993): 90–96.

Zimmer, Carl. "Location, Location, Location." *Discover* 14 (December 1994): 30–32.

The Interglacial

Anderson, P. M., et al. "Climatic Changes of the Last 18,000 Years: Observations and Model Simulations." *Science* 241 (August 26, 1988): 1043–1051.

Bower, Bruce. " 'Dwarf' Mammoths Outlived Last Ice Age." *Science News* 143 (March 27, 1993): 197.

Crowley, Thomas J. and Kwang-Yul Kim. "Milankovitch Forcing of the Last Interglacial Sea Level." *Science* 265 (September 9, 1994): 1566–1567.

Kerr, Richard A. "How Ice Age Climate Got the Shakes." *Science* 260 (May 14, 1993): 890–892.

Monastersky, Richard. "Deep Ice Stirs Debate on Climate Stability." *Science News* 144 (December 11, 1993): 390.

Neilson, Ronald P. "High-Resolution Climatic Analysis and Southwest Biogeography." *Science* 232 (April 4, 1986): 27–33.

Overpeck, Jonathan T., et al. "Climate Change in Circum-North Atlantic Region During Last Deglaciation." *Nature* 338 (April 13, 1989): 553–556.

Peltier, W. Richard. "Ice Age Paleotopography." *Science* 265 (July 8, 1994): 195–201.

Schnider, Stephen H. "Climate Modeling." *Scientific American* 256 (May 1987): 72–80.

Sullivan, Walter. "Great Climate Cycles Seen in Last Ice Age." *The New York Times* (February 1, 1994): C1 & C8.

White, J. W. C. "Don't Touch That Dial." *Nature* 364 (July 15, 1993): 186.

Zahn, Rainer. "Core Correlations." *Nature* 371 (September 22, 1994): 289–290.

Causes of Glaciation

Berner, Robert A. and Antonio C. Lasaga. "Modeling the Geochemical Carbon Cycle." *Scientific American* 260 (March 1989): 74–81.

Broecker, Wallace S. and George H. Denton. "What Drives Glacial Cycles?" *Scientific American* 262 (January 1990): 49–56.

Coffin, Millard F. and Olav Eldholm. "Large Igneous Provinces." *Scientific American* 269 (October 1993): 42–49.

Cordell, Bruce M. "Mars, Earth, and Ice." *Sky & Telescope* 72 (July 1986): 17–22.

Kerr, Richard A. "Did the Roof of the World Start an Ice Age?" *Science* 244 (June 23, 1989): 1441–1442.

Krumenaker, Larry. "In Ancient Climate, Orbital Chaos?" *Science* 263 (January 21, 1994): 323.

Kunzig, Robert. "Ice Cycles." *Discover* 10 (May 1989): 74–79.

Lo Presto, James C. "Looking Inside the Sun." *Astronomy* 17 (March 1989): 22–30.

Monastersky, Richard. "Rise of Tibet and Rockies Set Ice-Age Stage." *Science News* 135 (May 20, 1989): 309.

Oliwenstein, Lori. "Lava and Ice." *Discover* 13 (October 1992): 18.

Pearce, Fred. "Ice Ages: The Peat Bog Connection." *New Scientist* 144 (December 3, 1994): 18.

Pisias, Nicklas G. "Looking Ahead to the Milankovitch Anniversary." *Geotimes* 39 (November 1994): 7–8.

Ruddiman William F. and John E. Kutzbach. "Plateau Uplift and Climate Change." *Scientific American* 264 (March 1991): 66–74.

EFFECTS OF GLACIATION

Bower, Bruce. "Extinctions on Ice." *Science News* 132 (October 31, 1987): 284–285.

Boyle, Ed and Andrew Weaver. "Conveying Past Climates." *Nature* 372 (November 3, 1994): 41–42.

Crowley, Thomas J. and Gerald R. North. "Abrupt Climate Change and Extinction Events in Earth History." *Science* 240 (May 20, 1988): 996–1001.

Fischman, Joshua. "Flipping the Field." *Discover* 11 (May 1990): 28–29.

Garrett, Chris. "A Stirring Tale of Mixing." *Nature* 364 (August 19, 1993): 670–671.

Kerr, Richard A. "Ocean-in-a-Machine Starts Looking Like the Real Thing." *Science* 260 (April 2, 1993): 32–33.

Knauth, Paul. "Ancient Sea Water." *Nature* 362 (March 25, 1993): 290–291.

Kump, Lee. "Oceans of Change." Nature 361 (February 18, 1993): 592–593.

Monastersky, Richard. "Staggering Through the Ice Ages." *Science News* 146 (July 30, 1994): 74–76.

Stanley, Steven M. "Mass Extinctions in the Ocean." *Scientific American* 250 (June 1984): 64–72.

Vrba, Elisabeth S. "The Pulse That Produced Us." *Natural History* 102 (May 1993): 47–51.

Zimmer, Carl. "Inconstant Field." *Discover* 15 (February 1994): 26–27.

CONTINENTAL GLACIERS

Barnes-Svarney, Patricia. "Hubbard Glacier." *Earth Science* 40 (Fall 1987): 20.

Beard, Jonathan. "Glaciers on the Run." *Science 85* 6 (February 1985): 84.

Bromwich, David. "Ice Sheets and Sea Level." *Nature* 373 (January 5, 1995): 18–19.

Clark, Peter U. "Fast Glacier Flow over Soft Beds." *Science* 267 (January 6, 1995): 43.

Lehman, Scott. "Ice Sheets, Wayward Winds and Sea Change." *Nature* 365 (September 9, 1993): 108–109.

Maslin, Mark. "Waiting for the Polar Meltdown." *New Scientist* 139 (September 4, 1993): 36–41.

Monastersky, Richard. "Satellite Detects a Global Sea Rise." *Science News* 146 (December 10, 1994): 388.

Oliwenstein, Lori. "Cold Comfort." *Discover* 13 (August 1992): 18–20.

Peel, David A. "Cold Answers to Hot Issues." *Nature* 363 (June 3, 1993): 403–404.

Peltier, W. R. "Global Sea Level and Earth Rotation." *Science* 240 (May 13, 1988): 895–900.

Redfern, Martin. "Global Warming Cuts No Ice." *New Scientist* 139 (September 25, 1993): 16.

Stauffer, Bernard. "The Greenland Ice Core Project." *Science* 260 (June 18, 1993): 1766–1767.

Svitil, Kathy A. "The Sound and the Fury." *Discover* 16 (January 1995): 75.

THE ARCTIC

Cohn, Jeffrey P. "Gauging the Biological Impacts of the Greenhouse Effect." *BioScience* 39 (March 1989): 142–146.

Conover, R. J., et al. "Distribution of and Feeding by the Copepod Pseudocalanus under Fast Ice During the Arctic Spring." *Science* 232 (June 6, 1986): 1245–1247.

Gribbin, John. "Deepest Hole Yet over the North." *New Scientist* 139 (September 18, 1993): 18.

MacKenzie, Debora. "Did Northern Forests Stave Off Global Warming." *New Scientist* 139 (September 11, 1993): 6.

Monastersky, Richard. "Satellite Radar Keeps Tabs on Glacial Flow." *Science News* 144 (December 4, 1993): 373.

Pearce, Fred. "Does Polluted Air Keep the Arctic Cool?" *New Scientist* 144 (October 29, 1994): 19.

Pollack, Henry N. and David S. Chapman. "Underground Records of Changing Climate." *Scientific American* 268 (June 1993): 44–50.

Raney, R. Keith. "Probing Ice Sheets with Imaging Radar." *Science* 262 (December 3, 1993): 1521–1522.

Stone, Richard. "Signs of Wet Weather in the Polar Mesosphere." *Science* 253 (September 27, 1991): 1488.

Travis, John. "Taking a Bottom-to-Sky 'Slice' of the Arctic Ocean." *Science* 266 (December 23, 1994): 1947–1948.

Zimmer, Carl. "Son of Ozone Hole." *Discover* 14 (October 1993): 28–29.

The Ice Continent

Carpenter, Betsy. "Opening the Last Frontier." *U.S. News & World Report* (October 24, 1988): 64–66.

Eastman, Joseph T. and Arthur L. DeVries. "Antarctic Fishes." *Scientific American* 255 (November 1986): 106–114.

Gordon, Arnold L. and Josefino C. Comiso. "Polynyas in the Southern Ocean." *Scientific American* 258 (June 1988): 90–97.

Grotta, Daniel and Sally Grotta. "Antarctica: Whose Continent Is It Anyway?" *Popular Science* 240 (January 1992): 62–67 & 90–91.

Horgan, John. "Antarctic Meltdown." *Scientific American* 268 (March 1993): 19–28.

Jouzel, Jean. "Ice Cores North and South." *Nature* 372 (December 15, 1994): 612–613.

Kerr, Richard A. "Ocean Drilling Details Steps to an Icy World." *Science* 236 (May 22, 1987): 912–913.

Leal, Jose H. "Double-decked Ice Shelf." *Sea Frontiers* 39 (January-February 1993): 21.

Naeye, Robert. "The Strangest Volcano." *Discover* 15 (January 1994): 38.

Parfit, Michael. "Antarctic Meltdown." *Discover* 10 (September 1989): 39–47.

Radock, Uwe. "The Antarctic Ice." *Scientific American* 253 (August 1985): 98–105.

Glacial Structures

Brown, Barbara E. and John C. Ogden. "Coral Bleaching." *Scientific American* 268 (January 1993): 64–70.

Brown, Stuart F. "A New View of America." *Popular Science* 241 (November 1992): 86–89.

Chen, Ingfei. "Great Barrier Reef: A Youngster to the Core." *Science News* 138 (December 8, 1990): 367.

Fairbanks, Richard G. "Flip-flop End to Last Ice Age." *Nature* 362 (April 8, 1993): 495.

Folger, Tim. "The Biggest Flood." *Discover* 15 (January 1994): 37–38.

Goodwin, Bruce K. "The Hole Truth." *Earth Science* 41 (Summer 1988): 23–25.

Hoppe, Kathryn. "Bleaching Damage Spreads Beyond Corals." *Science News* 142 (November 14, 1992): 334.

Lipske, Mike. "Wonder Holes." *International Wildlife.* 20 (February 1990): 47–49.

Mollenhauer, Erik. "Glacier on the Move." *Earth Science* 41 (Spring 1988): 21–24.

Monastersky, Richard. "Coral's Chilling Tale." *Science News* 145 (February 19, 1994): 124–125.

——. "Spotting Erosion from Space." *Science News* 136 (July 22, 1989): 61.

Roberts, Leslie. "Greenhouse Role in Reef Stress Unproven." *Science* 253 (July 19, 1991): 258–259.

Glacial Deposits

Appenzeller, Tim. "After the Deluge." *Scientific American* 261 (December 1989): 22–26.

Barnes-Svarney, Patricia. "Beyond the Ice Sheet." *Earth Science* 39 (Summer 1986): 18–19.

Barron, Erik J. "Chill over the Cretaceous." *Nature* 370 (August 11, 1994): 415.

Hallet, Bernard and Jaakko Putkonen. "Surface Dating of Dynamic Landforms: Young Boulders on Aging Moraines." *Science* 265 (August 12, 1994): 937–940.

Kimber, Robert A. "A Glacier's Gift." *Audubon* 95 (May-June 1993): 52–53.

Krantz, William B., Kevin J. Gleason, and Nelson Cain. "Patterned Ground." *Scientific American* 259 (December 1988): 68–76.

Monastersky, Richard. "Stones Crush Standard Ice History." *Science News* 145 (January 1, 1994): 4.

——. "Hills Point to Catastrophic Ice Age Floods." *Science News* 136 (September 30, 1989): 213.

Moran, Joseph M., Ronald D. Stieglitz, and Donn P. Quigley. "Glacial Geology." *Earth Science* 41 (Winter 1988): 16–18.

Waters, Tom. "A Glacier was Here." *Earth* 4 (February 1995): 58–60.

Weisburd, Stefi. "Halos of Stone." *Science News* 127 (January 19, 1985): 42–44.

INDEX

Boldface page numbers indicate extensive treatment of a topic. *Italic* page numbers indicate illustrations or captions. Page numbers followed by *m* indicate maps, by *t* indicate tables, and by *g* indicate glossary.

A